Introduction to the General Principles of Aquaculture

Hans Ackefors, PhD
Jay V. Huner, PhD
Mark Konikoff, PhD

CRC Press
Taylor & Francis Group
Boca Raton London New York

CRC Press is an imprint of the
Taylor & Francis Group, an **informa** business

CRC Press
Taylor & Francis Group
6000 Broken Sound Parkway NW, Suite 300
Boca Raton, FL 33487-2742

First issued in paperback 2019

© 1994 by Taylor & Francis Group, LLC
CRC Press is an imprint of Taylor & Francis Group, an Informa business

No claim to original U.S. Government works

ISBN-13: 978-1-56022-012-1 (hbk)
ISBN-13: 978-0-367-40197-9 (pbk)

Library of Congress Cataloging-in-Publication Data

Ackefors, Hans.
    Introduction to the general principles of aquaculture / Hans Ackefors, Jay V. Huner, Mark Konikoff.
        p.   cm.
    Includes bibliographical references (p.   ) and index.
    ISBN 1-56022-012-0 (acid free paper).
    1. Aquaculture. I. Huner, Jay V. II. Konikoff, Mark. III. Title.

SH135.A25   1994   93-29833
639.8–dc20   CIP

Visit the Taylor & Francis Web site at
http://www.taylorandfrancis.com

and the CRC Press Web site at
http://www.crcpress.com

CONTENTS

ABOUT THE AUTHORS

Hans Ackefors, PhD, is Professor of Zoology and Head of the Division of Aquatic Ecology at the University of Stockholm in Sweden. He was chair of the Mariculture Committee of the International Council of the Exploration of the Sea from 1990-1993 and served as Chair of the Council for Planning and Coordination of Research in Sweden. Associate Editor of *Aquacultural Engineering,* the *International Journal of Aquaculture and Fisheries Technology,* and *Aquaculture in the Tropics,* he has published about 150 papers dealing with plankton, environmental control, fishery biology, and aquaculture. He is a member of the International Association of Astacology, the National Shellfisheries Association and the European Aquaculture Society, for which he is also past president.

Jay V. Huner, PhD, is Director of the Crawfish Research Center and Adjunct Professor of Aquaculture at the College of Applied Life Sciences at the University of Southwestern Louisiana in Lafayette. A Certified Fisheries Scientist, he has extensive teaching, research, and editorial experience in the field. Dr. Huner is the author of over one hundred technical and semi-technical fisheries publications and has co-authored or co-edited several aquaculture-related books. He is general manager for the International Association of Astacology (IAA), a member of the American Fisheries Society, the World Aquaculture Society, and the National Shellfisheries Association, and is active in many other professional organizations. Dr. Huner is editorially involved with the IAA Newsletter, *Farm Pond Harvest Magazine,* and the Louisiana Soft-Shell Crawfish Association Newsletter.

Mark Konikoff, PhD, is Associate Professor of Biology at the University of Southwest Louisiana and an Aquaculture Consultant for the U.S. Army Corps of Engineers Waterways Experiment Station. His research interests include aquaculture production systems, cage culture, polyculture, handling and hauling of live fishes, water quality, and ecology of fishes. He has published a number of articles and presented papers at professional meetings around the country. Dr. Konikoff is active in a variety of professional organizations, currently serving as President of the Introduced Fish Section of the American Fisheries Society.

List of Figures

List of Tables

List of Illustrations

Chapter 1

Introduction

"Aquaculture is the farming of aquatic organisms, including fishes, mollusks, crustaceans, and aquatic plants. Farming implies some form of intervention in the rearing process to enhance production, such as stocking, fertilizing, feeding, habitat manipulation, and protection from predators. Farming also implies individual or corporate ownership of the stock being cultivated" (FAO. 1991. Fisheries Circular 815 rev. 3).

Since the management of wild stocks of fishes taken by traditional fisheries may also involve similar enhancement techniques, the criterion of ownership is used by the Food and Agriculture Organization (FAO) to distinguish between aquaculture and fisheries harvests for statistical reporting purposes. Aquatic organisms that are harvested by an individual or corporate body who has owned them throughout their rearing period fall within the domain of aquaculture, while aquatic organisms that are exploitable by the public as a common property resource, with or without license requirements, are considered to be subject to the harvest of fisheries.

Incentives to cultivate aquatic organisms are now greater than ever. The natural harvest from the sea no longer increases by 4 to 5% annually as it did in recent decades. Overfishing and pollution have seriously reduced the quantity of seafood resources. The demand for many edible aquatic products exceeds the supply, especially the demand for high-priced "luxury" seafood. This has increased substantially in the industrialized world, exceeding the current supply of salmonids, flatfishes, seabreams, yellowtail, eels, lobsters, freshwater prawns, marine shrimps, freshwater crayfishes, and others. At the same time, the introduction of Economic Exclusive Zones (EEZ) in coastal waters during the 1970s has forced

1

many countries to look for alternatives to fishing to obtain food from the sea. Aquaculture seems to be the only way to increase supplies to meet increased demands.

In tropical and subtropical developing countries, aquaculture is more subsistence- than commercial-oriented because of the pressing need for animal protein in local diets. The production of fairly cheap aquatic products is simply a question of avoiding malnutrition and starvation. Fortunately, aquaculture in warm climates usually has an excellent potential for high production of fish and shellfish provided that a satisfactory water supply is available. Furthermore, since subsistence aquaculture has been a traditional practice in some parts of the world, such as Asia, and not traditional in others, such as the Near East and Africa, it could probably be easily transferred by applying an appropriate technology to these new areas.

Aquaculture is a very diversified business; its scope extends far beyond the production of fish and shellfish for the market. For example, aquatic farming also encompasses production of bait-fishes, ornamental fishes, special purpose fishes for biological control of weeds, mussels and macroalgae for therapeuticals and biochemicals, pearl oysters for pearls, reptiles for food and luxury leathers, and microalgae for fine chemicals and biogas. Finally, the controlled production of many types of aquatic test animals is within the scope of aquaculture.

Aquaculture has also become an integral component of fishery management programs. For the enhancement of fisheries in several countries, mass quantities of juveniles, especially finfishes, are cultured for stocking lakes and coastal areas. In some cases these fishes supplant those displaced by habitat alteration resulting from e.g., hydroelectric and navigation projects. Others are used as seed stock in new reservoirs and artificial ponds. Sometimes nonnative fishes are introduced to provide a new fishery. In a practice known as ocean ranching, anadromous (fishes that ascend freshwater streams to spawn) salmon smolts are cultured, released, and then subsequently recovered as adults at the release site. This, in turn, has led to increased catches of salmon on the high seas and coastal waters in both the Atlantic and Pacific oceans.

Aquaculture has an ancient history, with carp in China being the classic early example; but only recently has the dependence on natural production of seed for most species been overcome. A major reason for this advance has been the development of technologies to induce spawning by hormone injection; to feed the tiny larvae of marine fishes and crustacean species by supplying them with cultured live food such as microalgae, brine shrimp (*Artemia*), and rotifers; and to rapidly transport the cultured larvae in plastic bags. As a result, mass rearing of many of these valuable species is now possible.

The application of these new technologies, especially in saltwater, is still in its infancy. Therefore, despite recent advances in technology, the chief products of mariculture–the culturing of aquatic organisms in seawater–remain mollusks and seaweeds, usually cultivated by extensive and quite simple methods. In fact, most production of mussels and oysters is by farming. By comparison, less than 1% of all marine fish and 17% of crustacean harvests arise from mariculture. This is in sharp contrast to the situation in freshwater, where 60% of all fish production is cultivated. As the new technologies become more widely used, the contribution of mariculture to marine fish and crustacean catches should rise dramatically.

Aquaculture is a very complex business and much more difficult than conventional terrestrial agriculture. The purpose of this introductory text is to give readers an insight into the biological, technological, and economic factors that influence aquaculture production. The following questions are addressed: What is aquaculture? What are the prerequisites for and limitations of using an aquatic environment to cultivate aquatic biota? What is the present status of aquaculture on a global basis? What are the prospects for aquaculture in the future?

Usually people who become involved in aquaculture are very enthusiastic at the start, but it is best to combine this spirit with a realistic background of knowledge based on experience. We trust that this text will provide the reader with the basic knowledge needed to pursue a fascinating but complex endeavor, aquaculture–the farming of aquatic organisms.

Chapter 2

The History and Development
of Aquaculture

About 10,000 years ago, when people began to cultivate plants in addition to collecting seeds, nuts, and fruits, the first step was taken toward agriculture. At about the same time, wild animals were trapped and confined in enclosures. Gradually, the importance of hunting and gathering decreased as more and more plants and animals were domesticated. These developments, in turn, paralleled the growth of human populations and civilizations. Capture fisheries remain as the last major hunting and gathering activity and are still the foremost means by which we obtain aquatic animal protein from marine as well as freshwater environments. However, aquaculture, which has been practiced in various parts of the world for as long as 4,000 years, is becoming increasingly important.

The first known treatise on aquaculture, by the Chinese author Fan Li, dates from 475 B.C. The rearing of naturally spawning fish trapped in irrigation ditches was practiced in Japan more than 2,000 years ago. Stocking of carp into man-made enclosures was practiced during the first century, according to the Nihongi, the oldest written history of Japan. Tilapia farming is known from ancient Egypt, and the ancient Hawaiians and Mayans practiced aquaculture as well.

The practice of aquaculture was brought to Europe about 2,000 years ago. Aristotle mentioned carp in his papers; the Greeks and Romans fattened carp in ponds but they apparently confined their efforts to the trapping and on-growing of wild fish. The rearing of fishes through the entire life cycle developed much later in Europe, probably around the year 1150 A.D.

Carp farming was practiced in most European countries by the eighteenth century. Fish were kept in ponds to supply food during times when fresh meat was scarce. Monks were instrumental in disseminating fish farming methods because fish were eaten during the fast. Carp farming, however, was drastically reduced in many western European countries in the nineteenth century when sea fisheries developed and methods of seafood preservation and transportation were improved. In contrast to developments in western Europe, carp culture continues to be an important aquacultural endeavor in eastern European countries.

During the nineteenth century, American scientists started to breed rainbow trout for sport fisheries. The species was successfully transplanted from its indigenous range in the western part of North America to the eastern part of the continent, including the Great Lakes. A special seagoing strain of rainbow trout called the steelhead was also cultivated. Later, fertilized eggs were exported to Germany where culture of rainbow trout was started in freshwater ponds. The species was rapidly spread over many countries in Europe during the last two decades of the nineteenth century. Today, the rainbow trout is cultured all over the world as both a sport fish and directly for market.

Mariculture, too, has a long history. Rearing of marine clams in Japan dates from the eighth century. This practice was followed by seaweed farming 300 years later with transplantation of wild stock from one area to another. Farming of mussels and oysters commenced 800 years ago in France with collection of spats (settling larvae of mollusks) from wild spawn on a suitable substrate. In China where the alga *Porphyra* has been cultivated for at least 200 years, rocks were cleaned of other algae and barnacles to allow settling of wild sporophytes (the asexual part of the life cycle that produces the edible thallus).

The collection of wild seed (young organisms used to stock grow-out areas and/or ponds), its transportation, the preparation of suitable substrates, and the protection of the seed from competitors for food and space have always been and remain key steps in aquaculture. Important examples of this in Asian finfish culture are milkfish and yellowtail in the Philippines and Japan, respectively. The history of milkfish culture in the Philippines is over 300 years old. Wild seed is still collected in coastal areas and transplanted to aquacultural units in brackish mangrove areas. Collection of aquaculture seed from the wild, in fact, is a major industry in Asian countries, although it is now possible to induce spawning in many important species.

The controlled reproduction of marine fishes and shrimps is much more difficult than that of most freshwater and anadromous fishes. Although some marine fish species have been raised under experimental conditions for more than 100 years, mass production of juveniles has been a major bottleneck until recently. To overcome this problem, biologists in Europe formerly released newly hatched larvae for restocking purposes. An important example was cod larvae produced in specially designed hatcheries. Later, transplantation of wild caught seed was practiced. For example, millions of flatfish fry were caught in the North Sea and transplanted to areas with more favorable conditions. Both techniques were commonly used for many years between the First and Second World Wars. It was, however, never clearly demonstrated that these mass releases had any impact on the size of natural populations. In fact, the enthusiastic but ineffectual release of fry back into natural waters discredited the concept of aquaculture, and many people developed a less favorable opinion of this form of animal husbandry.

The mass rearing of marine fish and crustacean species from eggs has become a reality. In practically all cases, the larvae are very small, often less than 5 to 10 mm, and must be fed with living microorganisms which, in turn, must be cultured. Although by 1939 a technique had been developed to feed some marine fish larvae with nauplii of the brine shrimps, *Artemia*, there was no further progress in developing living larval feeds until the late 1940s. A major complication was the requirement for even smaller larval feeds by many species.

In Europe, it was shown experimentally that larvae of flatfishes like plaice, turbot, and sole could be raised on a diet of rotifers (a phylum of microscopic, mainly planktonic animals) and brine shrimp. Continuous study then led to commercial production systems for turbot and sole, and research is now underway to raise halibut. The main advancement in the rearing of marine finfishes in general, however, has been made in Japan during the course of the last several decades.

Aquaculture development has been especially spectacular in Japan. At least 1 million tons of seafood is now cultivated annually in coastal waters there. About 30 species of mollusks and 40 species of marine finfishes are currently mass-cultured in Japan with many more species being developed. The results of crustacean aquaculture are also impressive. Eight species of crab and 11 species of prawn and shrimp are mass-cultured either commercially or experimentally.

Before the 1950s, seed of many marine fish species were collected from the wild. In the 1960s, however, domesticated broodstock of many species were developed. Proficiency in the sophisticated technique of feeding microscopic organisms to larvae was the key to this success. The prevailing technique now consists of raising strains of green algae such as *Chlorella* as food for the rotifer *Branchionus*. The rotifers are then fed to the fish or crustacean larvae. Rotifers are also cultured on baker's yeast enriched with specific fatty acids to meet the particular dietary needs of a larval fish species.

No other country raises as many marine species for restocking purposes as does Japan. The excellent cooperation between fishermen and the staff of the hatcheries constitutes the foundation for this industry, which in a single year (1984) released nearly 30 million marine fishes, more than 300 million crustaceans, and close to two billion juvenile mollusks. The importance of restocking for commercial fisheries is considerable and may eventually become as important in Japan as conventional aquaculture.

A major event in the recent history of European aquaculture was the mass rearing of salmonids, especially the Atlantic salmon. During the 1940s and 1950s many hydropower stations were built on the rivers of Sweden and Finland. This resulted in the destruction of the natural spawning places for Atlantic salmon. Swedish law, however, forced the power companies to build hatcheries and produce salmon smolts to mitigate the loss of natural smolts. The combined efforts of power company and government/university aquaculturists has made it possible not only to maintain Atlantic salmon stocks in the Baltic Sea but also individual strains in specific rivers.

Atlantic salmon smolt production methods were also subsequently mastered in Norway by the mid-1960s. This was followed by grow-out studies in sea cages. The success of these studies is well-known, with Atlantic salmon aquaculture becoming a major commercial success not only in Norway but also throughout northwestern Europe.

Several notable aquacultural successes can be cited for North and South America. Examples in North America include Pacific salmon (both for food and for sustaining natural fisheries through stocking) in the U.S.A. and Canada, and rainbow trout, bait minnows, freshwater crayfish, oysters, ornamental fishes, and channel catfish in the U.S.A. The most important example in South America has been the development of pond culture of marine penaeid shrimp culture in Ecuador. There is also an increasing sea cage culture of salmonids in Chile. With the exceptions of rainbow trout and oyster culture, these endeavors are only two to three decades old.

The greatest aquaculture success in recent years has been the tremendous increase in world production of cultured marine penaeid shrimps. Pond production of penaeids grew from negligible amounts in the 1970s to 565,000 metric tons (mt), about 26% of the world harvest of shrimp in 1986. During the mid-1980s, shrimp culture had an average growth rate in excess of 32% annually. Shrimp farming is practiced in over 40 countries where an estimated total of 1.1 million hectares (2.7 million acres) is in production. Five countries–China, Indonesia, Thailand, the Philippines, and Ecuador–produce nearly 80% of cultured shrimp. The rapid growth rate of this industry can be attributed to several major breakthroughs in production technology, including nutrition, pond management, and especially hatchery production of larval and post-larval shrimp for stocking.

These historical examples of the development of aquaculture are discussed in greater detail later in this text. It seems appropriate, however, to complete this section where it began–in China. The development of Chinese polyculture with various carp species is indeed unique in the world. Many consider this polyculture of carps occupying different food niches to be an art based more on long experience with much trial and error than on scientific research. However, because annual production of fishes in China is now estimated at 4 million metric tons per year, this form of aquaculture is certainly the most important source of cultivated aquatic animal protein in the world. Chinese methods and species are being widely disseminated around the world, especially in developing countries.

Chapter 3

Natural Production versus Aquaculture

NATURAL AND ARTIFICIAL ECOSYSTEMS

Aquaculture includes the cultivation of aquatic organisms such as fish, shellfish, crustaceans, and plants using methods that increase the yield to a level above that naturally found in the environment. Thus, a natural food-producing ecosystem is changed to a more productive artificial, or manipulated ecosystem, in which a greater portion of the total energy input is used to increase yield per unit area.

Natural Aquatic Ecosystems–Fishing

Natural aquatic ecosystems are sun-based systems with relatively long food chains meshed in complicated food webs (see Figure 3.1). Generally, the energy transfer from one level on the food chain to the next (known as the gross ecological efficiency) is in the order of 10%. Thus, natural systems have high energy losses which ultimately result in a low yield per unit area. For example, overall biomass yield for the earth's oceans is less than 2 kilograms per hectare per year (kg/ha/yr).

In contrast to natural aquatic ecosystems, highly intensive artificial aquaculture ecosystems are characterized by high inputs of energy (feed/manure/heated water) and of seed, short food chains with low energy losses, and high (or higher) yields per unit area (see Figure 3.2). The range of production is on the order of 1,000 to 1,000,000 kg/ha/yr depending on the techniques and the energy input. Large amounts of wastes and nutrients can accumulate if not removed or dispersed. However, some aquacultural systems, espe-

FIGURE 3.1. Basic ecosystem components.

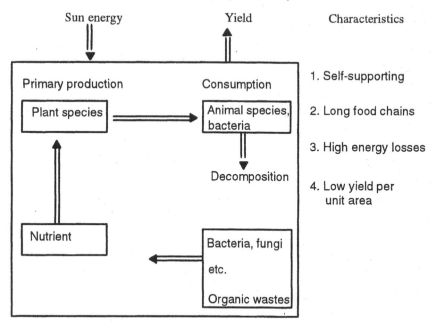

cially the most intensive, cannot be considered to be true ecosystems because they serve only one basic ecological function: consumption. All of the other natural functions usually performed by ecosystems such as decomposition of wastes, exchange of gases, and production of oxygen (by photosynthesis) are accomplished artificially by mechanical aeration and flushing.

DIFFERENT TECHNIQUES

Extensive, Semi-Intensive, and Intensive Aquaculture

The various levels of aquaculture progress from minimal to maximal inputs of external energy, ecosystem manipulation, and management. Extensive aquaculture involves a low degree of control over the environment, nutrition, predators, competition, and disease

FIGURE 3.2. Artificial "ecosystem"–aquaculture.

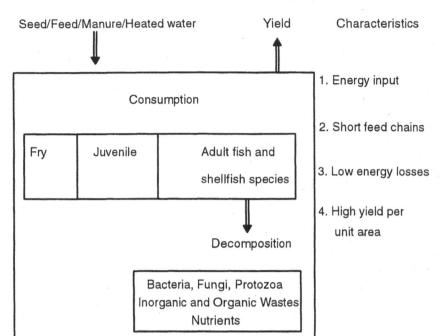

agents. Plant and animal seed stock is typically obtained from nature. Costs, technology, stocking rates, and production levels are all low. There is a high dependence on local climate and water quality. Natural water bodies and simple containment structures are used to culture the target organism(s). Natural food organisms, often generated within the cultural unit, sustain the system, although there may be occasional input of fertilizer, manures, or feedstuffs.

Intensive aquaculture involves a high degree of control over the system and high initial costs, a high level of technology, and high production levels. Seed stock is produced from domestic brood stock within the system, or is purchased from a specialized hatchery. There is little dependence on local climate and water quality. Man-made culture systems are used, including tanks, ponds, raceways, silos, cages, and so forth. There is a maximum output of product in a minimum of space and water. Virtually all nutrition comes from pelletized fabricated feeds which must be nutritionally complete. Water quality monitoring, mechanical aeration, and flushing are required on a daily basis.

Semi-intensive aquaculture involves a combination of some attributes of extensive and intensive aquaculture. It is usually done in man-made ponds. Costs, technology, stocking rates, and production levels are intermediate. Artificial feeds supply most of the nutrition of the cultured organisms in the pond, but their diet is supplemented with natural food organisms and these may make a significant contribution to essential nutritional requirements. Water quality is monitored daily, but waste removal and aeration are mainly accomplished by natural processes, even though these may be occasionally augmented mechanically. Examples of the various levels of aquaculture follow in the next section.

Examples of Aquaculture Techniques

There are many intermediate stages between highly intensive aquacultural systems and natural aquatic ecosystems from which man can harvest food. This is illustrated in Table 3.1. Very close to the natural ecosystem is the transplantation of seed from a nursery to a grow-out area using extensive farming practices. This technique involves culture of fish and shellfish species at low population densities with minimal use of supplemental feeds. Examples include artisanal culture of mullets, shrimps, and milkfish.

Most mussel and oyster cultivation involves extensive farming practices. Wild seed is collected on setting substrates. These are hard objects such as old mollusk shells on which the young mussels and oysters attach after they complete their planktonic larval stages. Although the attached mollusks may be moved and concentrated within a small area, only naturally produced plankton and detritus is utilized as the food source. Suspension of seed-covered substrates from rafts involves intensification, and some call this level of aquaculture semi-intensive, but it is still dependent upon nature for food and water.

Macroalgae are cultivated much the same way as the clinging bivalve mollusks. Thus, their culture is a form of extensive aquaculture with some intensification toward semi-intensive methods.

TABLE 3.1. Characteristics of natural ecosystems and man-made "ecosystems" with various techniques.

	Fisheries (temperate-tropical)	Extensive (subtropical)	Semi-intensive (temperate-tropical)	Intensive (temperate-tropical)
System type	1. Lake (temperate) 2. Lake (tropical) 3. North Sea 4. Rice field	1. Pond, polyculture 2. Lake, fisheries enhancement	1. Pond, polyculture 2. Pond, monoculture	1. Pond, monoculture 2. Cage, monoculture 3. Raceway 4. Silo
Source of productivity	1-4. Natural	1-2. Natural	1. Natural Fertilizer/feed 2. Feed	1-4. Fabricated feed
Food type	1-4. Natural	1-2. Natural	1. Natural/ Natural + Fertilizer/ Natural + Fertilizer + Pellets 2. Pellets	1-4. Pellets

Yield (kg · ha $^{-1}$ yr $^{-1}$)	1. 2-20
	2. 10-100
	3. 60-70
	4. 100-150
		1. 20-300
		2. 100-300
			1. 50-100/
			     200-300/
			     500-1,500
			
			2. 3,000-10,000
				1. 5000-100,000
				2. 100,000-300,000
				3. <1,000,000
				
				4. >1,000,000

Ocean (or lake) ranching and fisheries enhancement use more intensive forms of aquaculture during the initial seed production stages, but become extensive when the seed is released. Mature individuals in containers are used for spawning (natural or induced) to obtain seed. The young fish are usually raised from fry to juveniles before they are released in rivers, lakes, or sea areas for growout. They are subsequently captured in sport or commercial fisheries after an appropriate growth interval. The term "ranching" is used when taxa such as salmon smolts are released in a particular area and are recovered there after one or more years when they return as adults to spawn.

Semi-intensive culture of fish or shellfish is usually practiced in ponds. Manure, sewage, and/or supplementary feed are provided. This means that the fish or shellfish may utilize both natural and fabricated feeds. Tilapia, carps, catfishes, shrimps, and prawns are usually raised in this way (see Illustration 3.1).

ILLUSTRATION 3.1. Rice-cum-fish culture in the Orient. Deep trenches in field provide refuges for hardy fish like tilapias during the heat of the day and they forage in shallow water at night. Credit: J. W. Avault, Jr., Louisiana Agricultural Experiment Station.

Integrated aquaculture-agriculture methods are well developed in Southeast Asia. Geese, ducks, chickens, and/or pigs are commonly raised with various fish species. This is the so-called animal-cum-fish technique (see Illustration 3.2). Wasted feeds and manures provide sustenance and fertilizer for polyculture systems with many different carp species. These various carp species utilize different pond niches and this permits the aquaculturist to maximize fish production. These systems are thus semi-intensive, even though annual productivity in them can be several times greater than the 3,000 to 5,000 kg/ha productivity in some forms of intensive monoculture such as catfish in North America. Although these traditional integrated duck-cum-swine-cum-fish ponds have great efficiency, there has been recent speculation that the extreme proximity of species, especially ducks, swine, and humans, combined with the recycling of untreated wastes, has led to the periodic development

ILLUSTRATION 3.2. Duck-cum-fish culture in Philippines. Duck droppings and wasted feed "feed" low tropic level fishes. This form of culture is common in the Orient and Eastern Europe. Credit: J. W. Avault, Jr., Louisiana Agricultural Experiment Station.

of new strains of human pathogens (mainly flu viruses) in these systems.

Intensive aquaculture systems involve a range of technological levels and may include ponds, tanks, raceways, or cages and also high technologies such as recirculating systems or flow-through systems with heated water. Fabricated l feed is used exclusively and the yield per unit area is high (see Illustrations 3.3-3.9). Intensive culture usually implies that the level of feeding is such that some form of artificial life support system, such as aeration, is required. In this definition, most of the commercial catfish ponds in the U.S.A. would be classified as intensive. Compared to a fishery that harvests the natural production in a body of water, the annual yield of intensive aquaculture is 10 to 10,000 times higher, depending on the input of energy and the methods employed.

ILLUSTRATION 3.3. Historical earthen rainbow trout raceways in Denmark. Note demand feeders in each pond. Credit: Jay Huner.

ILLUSTRATION 3.4. Primitive floating cages in Hong Kong used to culture grouper and various seabreams. Structures house guards. Credit: Jay Huner.

ILLUSTRATION 3.5. Modern intensive culture pools for rainbow trout with hatchery building in background and earthen raceways in foreground. Credit: EWOS AB, Södertälje, Sweden.

ILLUSTRATION 3.6. Barge-like hexagonal floating cage support structure. Boom is used to deliver feed from central control structure and move harvested fish to it. Credit: Jay Huner.

ILLUSTRATION 3.7. Large circular carp egg incubation/hatching pools in southern China. Millions of eggs can be handled in a very small area. Credit: Jay Huner.

ILLUSTRATION 3.8. Hall used to cultivate fingerling trout and salmon in Finland. Credit: Jay Huner.

ILLUSTRATION 3.9. Enclosed fish processing/hatchery unit in central Finland. Note large feed storage tanks to the right, tractor for dispensing feed to the left, and large earthen raceways in foreground. Credit: Jay Huner.

Chapter 4

The Environment for Aquatic Organisms

AIR VERSUS WATER AS CULTURING MEDIA FOR BIOTA

Air and water are dramatically different media. Water is very dense, and so favors bulky organisms by supporting them and thus minimizing their energy expenditures. The greater viscosity of water is also used to advantage by many aquatic organisms as an aid for locomotion. This combination of high density and viscosity is what allows most fishes and many aquatic invertebrates to swim efficiently through water, while most land animals are limited to what is essentially crawling over the "bottom" for moving about. Water also differs from air in that it has a much greater heat capacity. This results in temperatures being more stable in water, changing much more slowly than they do in air. Therefore, aquatic organisms generally have not had to develop complex physiological systems to compensate for dramatic changes in temperatures as have organisms dwelling on the land.

Air and water differ in oxygen content. Oxygen is never a limiting factor to crops cultivated on land but is frequently a limiting factor in aquatic culture systems. Very little oxygen enters the water from the atmosphere via diffusion. Therefore, the principal source of oxygen is photosynthesis. But even clear water quickly absorbs solar radiation, and so photosynthesis is limited by the depth to which solar radiation penetrates into an aquatic system–that is, to surface layers.

A final difference between the two media is the amount of water physiologically available to organisms. Those dwelling in an aqueous medium never face the problems of drought and desiccation associated with land; however, osmotic stresses are much more important to aquatic organisms than to those living or being cultivated on land.

CHARACTERISTICS OF WATER

The special chemical and physical properties of water make aquatic and terrestrial life possible on earth. The specific structure of the H_2O molecule makes water a unique liquid with high density, high melting point, high heat capacity, high boiling point, high steam generation point, and high surface tension.

Some properties of water are of particular importance for aquaculture: (1) the high density enables organisms to remain in suspension and to move with minimal expenditure of energy, (2) the solubility of gases such as oxygen satisfies the respiration needs of aquatic organisms, and (3) the flow and dilution capacity of water permits organisms to produce very simple but highly toxic excretory products such as ammonia which are readily diluted and carried away. Thus, aquatic organisms avoid the complex transformations and concentration of nitrogenous wastes that terrestrial animals must perform.

The quantity and quality of water are the most important considerations in developing an aquacultural endeavor. Competition for water from other activities—recreational, industrial, agricultural, and/or shipping—seriously hampers or even precludes aquacultural developments in areas with otherwise excellent potential.

The amount of water required for farming specific species is determined initially by that species' oxygen demand. This depends on its developmental stage, activity level, stocking density, temperature, and feeding rate. Active cold-water species such as the salmonids require oxygen concentrations at or above 5 to 6 mg O_2/l (milligrams of dissolved oxygen per liter). Oxygen-rich water must be constantly renewed because the fish continuously consume oxygen and release carbon dioxide. Table 4.1 provides information on the flow rates required to provide sufficient dissolved oxygen and remove wastes in a fixed raceway cultural unit. Water needs are functions of fish size and temperature. Small fishes have a higher metabolic rate and therefore consume more oxygen per unit weight than do larger fish of the same species. Metabolic rate also increases in warmer water.

The tabular data show clearly that the amount of water required decreases with increasing size (and age) and decreasing water tem-

TABLE 4.1. The amount of water (95% saturated with dissolved oxygen) required in intensive salmon culture systems as a function of fish size and temperature. Liters of water per minute and per kg fish.

Individual fish	Temperature			
weight (g)	4° C	10° C	14° C	18° C
0.1	0.4	1.2	1.8	2.8
1.0	0.3	0.8	1.2	1.8
10	0.2	0.5	0.8	1.2
100	0.1	0.3	0.5	0.8
1,000	0.05	0.2	0.3	0.5

peratures. Chemical characteristics of the water, such as pH, nitrite and ammonia concentrations, and alkalinity, among others, are also important, so that the tabular oxygen consumption values can only be looked upon as general guides in determining the amount of water required to culture the fish.

By using various techniques, it is possible to decrease the amount of water needed to provide adequate dissolved oxygen to the fish. The water can be oxygenated by simple mechanical stirring/agitation or by adding gaseous or liquid oxygen. Table 4.2 provides some overall figures for the amounts of water used in three different systems for culturing warm and cold water fishes. These data show that water use (and oxygen requirements) increase with the intensity of management. For example, the stocking density is greatest in the Japanese system. As a result, the amount of fish produced per unit area per year is about 2,000 times higher than that in the standing water pond in Israel. However, Israel is an arid land where water is scarce and a Japanese flow-through system is strictly out of the question there.

It is certainly clear, then, that huge amounts of water are necessary for culturing most freshwater species. This is especially important in planning aquacultural development. For example, in Hawaii, freshwater flows in grow-out ponds for various species are 93 to

TABLE 4.2. Amount of freshwater used per ton of cultured fish in three different aquaculture systems.

System	Location	Species	m³ water/ton fish
1. A pond with flow-through system	Takasaki City, Japan	Carp	740,000
2. Raceway	Idaho, USA	Trout	210,000
3. A pond with stagnant water	Dor, Israel	Polyculture (Carp, mullet, tilapia)	6,000-8,000

271 liters per minute per hectare. The forecast for expansion of aquaculture in the Hawaiian Archipelago for the year 2000 shows that 6,000 ha of ponds will then be used for aquacultural endeavors, and the amount of freshwater needed will be equivalent to that used by a city of 2.0 to 5.8 million people.

Such a comparison with human needs may, however, be improper if it does not include a consideration of competitive uses of water by humankind. Indeed, industry and agriculture are the two primary consumers of freshwater. More than 3,000 liters per person per day are utilized in the U.S.A. for industrial purposes. In contrast, the average industrial requirement for water on a worldwide basis has been established at 1,200 liters per person per day. Multiplied by five billion people and 365 days, this generates a most impressive volume of water. Table 4.3 provides reference data for the volumes of water needed to produce various industrial products.

Agriculture consumes its share of water, too. About half of the agricultural production in the world is dependent on irrigation, and 2,500 km³ $(2.5 \times 10^{15}$ liters) of water is used per year for this purpose—roughly the same order of magnitude as the amount required by industry. While the basic daily requirement of water for personal needs has been estimated at only 3 liters per person world-

TABLE 4.3. Water requirements for various industrial processes.

Product	Water Used in Production (liter per kg)
Nitrogen fertilizer	600
Sugar	100
Steel	150
Artificial silk	1,000
Paper	250
Oil	180
Brick	2
Plastic	750-2,000

wide, people in industrialized countries each consume 400 liters per day, mainly for hygienic activities. Assuming a worldwide average daily use of 100 liters per person, the world's domestic water requirements amount to about 180 km^3 (2×10^{14} liters) per year. This is only 7% of the amount used for irrigation and 8% of the amount used by industry. Thus, the volumes of water necessary for freshwater aquaculture are best compared with those utilized by industry and agriculture.

Several aquacultural technologies have been developed to make optimal use of limited water resources. There are basically two types of aquaculture systems in terms of water use: (1) standing water systems and (2) flowing water systems. The flowing water systems can be further divided into two types: (1) flow-through systems and (2) recirculation systems. Reuse of water decreases the total amount of water needed per unit of fish produced but with several additional costs, including (1) pumping, (2) aerating, and (3) cleansing the water. This requires a higher initial investment and the use of more energy than that required for flow-through systems. However, there are situations where recirculation is cost effective, such as in cold climates where the use of heated water is a must and reuse of water conserves "expensive" heat.

Table 4.4 compares different farming systems in terms of mean residence time of water. Mean residence time is the period that an

TABLE 4.4. Comparison of water exchange in four aquaculture systems.

Type of system	Theoretical water exchange per day	Rearing type
1. Open	25-100	Net cage
2. Open Aeration	10-15	Basin with water pumps
3. Semi-closed Water treatment Oxygenated water	0.5-4	Recirculation system e.g., a hatchery or a nursery unit
4. Closed Water treatment Oxygenated water	0.0-0.5	Experimental recirculating system

average water molecule is expected to remain in the culture unit before being washed out. The reciprocal of the mean residence time is the theoretical number of water exchanges. For example, if the mean residence time were six hours (0.25 days) then there would be 4.0 theoretical water exchanges per day. The data for the theoretical daily water exchanges are based on a stocking density of 1 kg of fish per 25 to 75 liters of water. The first two systems are open (flow-through), whereas the remaining two systems are semiclosed or closed (recirculating). The mean residence time of water increases from the first system to the last system.

Most flow-through systems require enormous amounts of water. Some types of flow-through systems, however, have a very low turnover of water and come close to standing water systems. In land-based systems, such as raceways and tanks, where water must be pumped through the culture container, every effort is made to reduce the amount of water needed. This necessitates aeration to restore oxygen used by the fish or shellfish in the system. In the third system, the theoretical number of daily water exchanges is relatively low. In such semiclosed systems, the water must be circulated through a settling basin, a biofilter, an oxygenation unit, and other water treatment systems to remove the bulk of the waste materials from it before it returns to the culture vessel. The most extreme form of recirculation is the completely closed system in which only evaporated water is replaced; but there, water purification systems must be capable of totally removing all wastes produced.

WATER QUALITY

The term "water quality" in the context of aquaculture generally refers to all factors–physical, chemical and biological–that influence the well-being of aquatic biota. The term "quality" implies that no factor should exceed certain upper limits for toxic compounds, or fail to remain within some minimum-maximum range for life sustaining physico-chemical factors such as pH, dissolved oxygen, temperature, etc.

Water quality factors can affect aquatic species both negatively and positively. For example, higher temperatures, within limits, increase growth rates because metabolism increases, but this also increases susceptibility to toxic materials such as pesticides. The major physical and chemical water quality factors that impact aquaculture are discussed in this section.

Temperature

Most aquaculture species are poikilothermic, meaning that their body temperatures conform to the temperatures of their environments. They can, however, seek out favorable thermal regimes. All vital functions–basal metabolism, growth, reproduction, etc.–depend on body temperature. Therefore, the rates of all these functions increase or decrease according to variations in the surrounding water temperature. Rate changes are affected by the so-called Q_{10} effect: for every 10°C of temperature change, rates double or half, depending on the direction of the change (that is, provided the change is within the species' genetically predisposed and environmentally modified tolerance limits). Too rapid or too great a change may be deadly because the organisms are unable to accommodate metabolically. However, in natural situations, temperatures usually change gradually enough for organisms to adjust metabolic rates in an orderly manner.

Susceptibility to diseases, parasites, and nonliving toxicants is greatly affected by temperature. Fish and shellfish are often referred to as cold, cool, and warmwater species depending on the thermal regime to which they are naturally adapted. Trouts and salmons are coldwater species. These cannot survive sustained temperatures above 20°C. Percids like yellow perch and pike perch are coolwater species. These thrive at temperatures about 20°C, but can tolerate both warmer and colder temperatures. Catfishes and largemouth bass are typical warmwater species. They do best at temperatures of 25 to 30°C, but can survive both warmer and much colder temperatures. The so-called tropical fishes such as tilapia and gouramis flourish at temperatures near 30°C, and will die if temperatures get below 15 to 20°C, depending on the species. For further information see Table 4.5.

Thermal optima (ideal temperatures) may vary according to life cycle stage, and it is essential for successful aquaculture to be aware of these optima. Most optimum temperatures presented in the literature refer to those that provide for the best growth rate. However, the best feed efficiency ratios (gain/feed) are usually obtained at temperatures lower than those associated with optimal growth rate. The data in Table 4.5 clearly show that there are significant differences in optimum temperatures for various fishes. Thus, the aquaculturist will have to seek to provide the optimum temperature, but, at the same time, balance this with the thermal regime that provides for the best feed efficiency ratio while reducing the species sensitivity to various diseases. Although, in principle, the higher temperatures are more conducive to the development of pathogens, colder water (in the autumn) can also make species like the European crayfish more sensitive to infection by making the immune system less responsive to challenges.

Extremely low winter temperatures in high latitudes or altitudes can also cause serious problems in some marine and freshwater systems. The presence of dissolved salts in marine waters lowers the freezing point below 0°C and such waters can become super-

TABLE 4.5. Mean optimum temperatures for growth of some fish species.

Common Name	Scientific Name	° C
Brown trout	Salmo trutta	13
Chum salmon	Oncorhynchus keta	13
Brook trout	Salvelinus fontinalis	14
Arctic char	Salvelinus alpinus	14
Atlantic salmon	Salmo salar	14
Sockeye salmon	Oncorhynchus nerka	15
Chinook salmon	Oncorhynchus tshawytscha	16
Pink salmon	Oncorhynchus gorbuscha	16
Rainbow trout	Oncorhynchus mykiss	17
White fish	Coregonus artedii	18
Eel	Anguilla anguilla	23-25
Carp	Cyprinus carpio	23-27
Largemouth bass	Micropterus salmoides	25-27
Channel catfish	Ictalurus punctatus	30

cooled. This can be disastrous in cage/pen units in coastal waters. The freezing point (T) is around -0.5 to $-1.5°C$ at salinities ranging from 10 to 30 ppt (parts per thousand) (T $= -0.054S$; where S = salinity in parts per thousand or ‰). Such temperatures are lethal to, for example, salmonids. Caged fish exposed to these temperatures cannot move to warmer water and will suffer heavy mortalities.

Dissolved Gases

The amount of oxygen, carbon dioxide, and nitrogen that can dissolve in water is an inverse function of water temperature and salinity. It is also affected by air pressure, total gas pressure, and water depth. Table 4.6 shows the solubility of oxygen at various temperatures and salinities. The importance of dissolved oxygen (DO) cannot be overstressed. Only specialized aquatic organisms can exist anaerobically and few or none of these have aquaculture potential. While some cultured species such as snakeheads and crayfishes can literally breathe air when dissolved oxygen is low

TABLE 4.6. The solubility of oxygen in mg/l as a function of temperature and salinity at an air pressure of 1013 mbar (760 mm Hg).

Temperature	Salinity							
	0	5	10	15	20	25	30	35
0° C	14.6	14.1	13.6	13.2	12.7	12.3	11.9	11.5
5° C	12.8	12.3	11.9	11.6	11.2	10.8	10.5	10.1
10° C	11.3	10.9	10.6	10.3	9.9	9.6	9.3	8.9
15° C	10.1	9.8	9.5	9.2	8.9	8.6	8.4	8.1
20° C	9.1	8.8	8.6	8.3	8.1	7.8	7.6	7.3
25° C	8.2	8.0	7.8	7.6	7.4	7.2	7.0	6.8
30° C	7.5	7.3	7.1	6.9	6.8	6.6	6.4	6.2

(hypoxic conditions), this puts them under stress that affects all body functions. Air contains 210,000 mg/l of oxygen. Pure water seldom contains more than 10 mg/l of dissolved oxygen. Thus, aquatic species must be able to complete their life cycles in systems with limited oxygen levels compared to those available to terrestrial organisms. One episode of oxygen depletion will destroy an entire aquaculture crop. Thus, maintaining adequate oxygen levels represents the aquaculturist's most pressing priority.

Table 4.6 shows that at saturation, freshwater contains substantially more oxygen than normal seawater (35 ‰ or parts per thousand–salinity). Secondly, it shows that the yearly seasonal temperature changes in temperate climates have a marked influence on dissolved oxygen. This inverse relationship between oxygen solubility and temperature is particularly challenging to aquatic organisms because increased temperatures result in an increased metabolic rate and greater need for oxygen while simultaneously reducing the concentration of dissolved oxygen available.

The concept of "oxygen saturation" must be interpreted with caution. For example, in pure, saturated freshwater there are 12.8 mg/l of oxygen at 5°C while at 25°C there are 8.2 mg/l, 4.6 mg/l less than at 5°C. However, it must be emphasized that all aquatic species function best when water, regardless of temperature and salinity, is nearly saturated with oxygen. Thus, aquaculturists need to manage their systems to hold oxygen levels as near saturation as possible.

The importance of dissolved oxygen simply cannot be overemphasized. Oxygen saturation should never be below 90% for culturing salmonids and the oxygen concentration in the effluent (usedwater exiting the raceway or culture unit) should not fall below 5 mg/l. Warmwater fish species such as carp are naturally more tolerant of low oxygen levels because warm water holds less oxygen than cold water. However, oxygen levels in carp culture systems should not fall below 3 mg/l to realize maximum growth rate and feed conversion ratios.

In static and semi-static ponds, the daily fluctuations of dissolved oxygen must be closely monitored. During daylight hours, oxygen supersaturation often occurs, especially in smaller, organically enriched units, because of intensive photosynthesis by the phytoplankton. Conversely, oxygen can be almost totally depleted during the night because of respiration of all biota present, including the cultured species, macrophytes, phytoplankton, zooplankton, and bacteria living in either the water column or in the bottom sediments.

Dissolved carbon dioxide, originating from diffusion from the atmosphere and respiration of aquatic biota, is another important gas. Carbon dioxide reacts with water to produce carbonic acid. The reactions of dissolved carbon dioxide are as follows:

$$CO_2 + H_2O \rightleftharpoons H_2CO_3$$
$$H_2CO_3 \rightleftharpoons H^+ + HCO_3^-$$
$$HCO_3^- \rightleftharpoons H^+ + CO_3^=$$

Thus, carbon dioxide is present in water in various chemical forms (CO_2, H_2CO_3, HCO_3^-, and $CO_3^=$), and the relative amount of each compound is related to the pH of the water and, to a lesser extent, temperature. Table 4.7 shows the key role of pH on the shifts of carbon dioxide from one form to the other in water. The tabular data show that carbon dioxide is mainly present in acid waters as CO_2

TABLE 4.7. Percentages of carbon dioxide + carbonic acid, bicarbonate, and carbonate in freshwater at various pH values at 8° C.

pH	$CO_2 + H_2CO_3$	HCO_3^-	$CO_3^=$
5.0	69.9	3.1	0
6.0	75.8	24.2	0
7.0	23.6	76.4	0
8.0	3.0	96.7	0.3
9.0	0	96.7	3.0
10.0	0	76.9	23.1

and the un-dissociated acid H_2CO_3. In neutral and slightly alkaline waters, bicarbonate, HCO_3^-, is the major form and in very alkaline waters, the bicarbonate is dissociated further to carbonate, CO_3^{2-}.

In most natural waters, the carbon dioxide concentration seldom exceeds 6 mg/l. The amount is regulated by biological factors (respiration and bacterial degradation of organic material) and physical factors (pH, temperature, and carbon dioxide partial pressure in the atmosphere). It should be emphasized that, in comparison to the few mg/l of carbon dioxide in surface waters, ground water may contain a hundred times more carbon dioxide. When using such water for aquacultural purposes, it is necessary to first remove the excess carbon dioxide, usually by aeration.

The amount of carbon dioxide in water can reach higher than normal levels in aquaculture operations and should be kept under control. Indeed, a high amount of dissolved carbon dioxide may prevent complete saturation of oxygen in hemoglobin and cause respiratory stress. As seen in Table 4.8, fishes have different sensitivities to increased concentrations of carbonic acid.

Observations have shown that in nature, fishes tend to avoid waters with a carbon dioxide level exceeding 1.6 mg/l. In intensive rearing systems, the amount of this compound may, however, increase to values ten times higher. This results from accumulation of respired carbon dioxide from the cultured species as well as from bacterial respiration of organic wastes.

TABLE 4.8. Effects of carbonic acid (mg per liter) on different fishes.

Fish species	Observed effects			
	1. Restless-ness	2. Drowsi-ness	3. Dizzi-ness	4. Lying on side
Rainbow trout	9-18	36	36-73	55-73
Brook trout	9-18	36	35-55	46-64
Carp	18-36	55-73	202	239

Water can become supersaturated with a variety of dissolved gases (but especially nitrogen), any of which can generate a syndrome called "gas bubble disease." Gas bubble disease occurs when the water holds more gas than it should for its temperature. Supersaturation is a temporary condition and the water returns to equilibrium as the excess gas effervesces to the atmosphere. However, when fish and shellfish are exposed to supersaturated water, they absorb the gas and it is released from suspension in body fluids forming potentially life threatening internal bubbles.

An example involves supersaturation with nitrogen of cooling waters from power stations. If this water is used as a direct heat resource in a fish culture unit it can cause gas bubble disease. The water is supersaturated with nitrogen because it has been rapidly heated in the power plant and the amount of gas still present in the outflow exceeds the normal saturation value. Supersaturation can also occur when water is subjected to sudden pressure changes, such as being pumped from some deep wells, from faulty water supply systems that allow the injection of air into water flowing in enclosed pipes, and from waterfalls and hydropower stations.

Supersaturated water is usually only a problem in situations where the fish are confined in raceways, tanks, aquaria, or cages, and are continually supplied with the gassy water. It is usually not a problem in ponds because any excess gas in the water supply is released from the pond surface before it can build up. In some aquaculture systems, especially ponds, the water may also be slightly supersaturated with oxygen, but this is usually not a problem because there is no persistence of supersaturation in the cultural unit.

Nitrogen and Phosphorus Compounds

Lakes and seas receive nitrogen and phosphorus compounds from runoff and river inputs carrying wastes from agricultural, industrial, and other human activities. Organic compounds are mineralized by bacteria, forming ammonia, nitrites, nitrates, and phosphates. Excessive amounts of these nutrients ultimately lead to the well-known phenomenon of eutrophication, which has become a serious problem in many inland and coastal aquatic ecosystems. Inland water bodies can be classified according to the level of dissolved nutrients that they contain. At one extreme are crystal-clear, low-nutrient lakes that have low annual productivity in terms of kg/ha, but that have high oxygen levels in all seasons, even in deep water. These are known as oligotrophic lakes and they provide habitat for such sensitive fishes as ciscos, whitefishes, lake trout, and other salmonids. In contrast, lakes that have high nutrient levels and high annual productivity are more turbid and usually do not have dissolved oxygen in the deeper waters during mid-summer. These are known as eutrophic lakes and produce such fishes as catfishes, carps, sunfish, and black bass. The oligotrophic lakes gradually become more fertile and transform into eutrophic lakes, but this natural process of eutrophication can take thousands of years. The effect of human activities is to greatly accelerate the eutrophication process and thus endanger populations of fishes especially adapted to low-nutrient oligotrophic conditions. Further enrichment of already eutrophic lakes can also result in excessive plankton blooms and periodic fish die-offs due to low oxygen.

In aquacultural operations, the metabolites produced by the cultivated species and their food residues contribute measurably to enrichment of waters receiving system effluents. The main excretory product of most fishes is ammonia, which is discharged into the water through the gills. In addition, dissolved urea as well as particulate wastes (feed and feces) in the water are converted chemically and/or microbiologically to inorganic compounds such as ammonia and phosphate. The ammonia is further converted first to nitrite and then to nitrate.

Some of these compounds, like free ammonia and nitrite, are very often toxic to aquatic biota and their accumulation should be avoided. In acidic conditions, free ammonia (also called un-ionized ammonia) is transformed into the much less toxic ammonium ion (NH_4^+). Therefore, the pH of the culture medium is of paramount importance.

$$NH_3 + H_3O \rightleftharpoons NH_4^+ + H_2O$$

This reaction is influenced by temperature and salinity as well. Tables 4.9 and 4.10 provide the percentages of the toxic NH_3 in freshwater and seawater respectively, at different temperatures and pH's. As can be seen from both tables, the amount of toxic free ammonia increases rapidly as pH increases as well as at higher

TABLE 4.9. Percentage of un-ionized ammonia (NH_3) in freshwater at various temperatures and pH's.

Temp./pH (in ° C)	6.0	7.0	8.0	9.0	10.0
5	0.0125	0.125	1.23	11.1	55.6
10	0.0186	0.186	1.83	15.7	65.1
15	0.0274	0.273	2.67	21.5	73.3
20	0.0397	0.396	3.82	28.4	79.9
25	0.0569	0.566	5.38	36.3	85.1

TABLE 4.10. Percent of un-ionized ammonia in seawater (salinity = 23-27 ‰) at different temperatures and pH's.

Temp./pH (° C)	7.5	8.0	8.5
10	0.492	1.54	4.71
15	0.713	2.22	6.70
20	1.03	3.19	9.44
25	1.49	4.57	13.10

water temperatures. Comparison of the two tables also shows that, under the same temperature and pH conditions, the percentage of un-ionized ammonia is lower in seawater than in freshwater. This is due to the fact that the ionic strength of the medium also influences the transformation process.

In natural waters, ammonia is converted rather rapidly to nitrite and further to nitrate by bacteria of the genera *Nitrosomonas* and *Nitrobacter*. Accumulation of these end products of the biological cycle is, however, counteracted in natural ecosystems by the perpetual synthesis of organic matter through photosynthesis; that is, the uptake of nitrates and phosphates by photosynthesizing plants and algae.

Ammonia and nitrite are the most toxic self-generated compounds in rearing units and the literature is replete with recommended values that should not be exceeded under particular conditions. Unfortunately, these values are often given in different units: ammonia-nitrogen (mg/l $NH_3 - N$) or free ammonia (mg/l NH_3) concentrations. The $NH_3 - N$ values can be converted into free ammonia concentrations by multiplying by a factor of 1.22. However, the actual percentage of total ammonia ($NH_3 + NH_4^+$) that is in the un-ionized (toxic) form (NH_3) is dependent on temperature, salinity, and pH (see Tables 4.9 and 4.10).

In order to be on the safe side with regard to toxic effects from nitrogen compounds, the following concentrations of free ammonia, nitrite, and nitrate (in mg/l) in rearing units should not be exceeded over extended periods of time: NH_3, 0.005; NO_2^-, 0.01; and NO_3^-, 50-100. For short-term exposures, concentrations may be five to 50 times higher without any apparent adverse effects. As a general guide, maximum short-term concentrations for NH_3, NH_4^+, and NO_2^- are 0.08 mg/l, 3.0 mg/l, and 0.5 mg/l, respectively. The toxicity of nitrite to freshwater fishes is greatly affected by the presence of chloride ions, with maximum protection coming when there is a 4:1 ratio between chloride and nitrite. Therefore, temporary elevations of nitrite are often treated by adding an appropriate amount of NaCl to the culture water. This is commonly done in catfish ponds in the U.S.A.

pH and Alkalinity

Most freshwater species live in waters with a pH range of 5.0 to 9.0. A pH value indicates the amount of positive hydrogen ions present, with values of 0 to 7.0 being acidic, 7.0 being neutral, and 7.0 to 14.0 being basic (alkaline). Some organisms survive at pH 4.0, whereas others cannot tolerate values lower than 6.0 to 6.5. Spawning and development of eggs and fry of most fishes are usually affected by pH values lower than 5.5 to 6.0, whereas adults can tolerate such acid waters. In neutral water (pH = 7.0) oxygen consumption is lowest; respiration increases at lower and higher pH values.

Changes in pH in natural waters are mainly caused by the cyclic photosynthetic activity of aquatic plants (phytoplankton, nonvascular plants, and vascular plants) and by the respiration of all aquatic biota over each 24-hour period. These two activities have opposite effects on the pH. The pH is lowest in the early morning before the sun rises, when the respiration is greatest, and highest in the late afternoon before the sun sets, when photosynthesis is greatest. This cycle is especially apparent in ponds with little water flow. In photosynthesis, carbon dioxide is removed from the water and converted into carbohydrate with formation of oxygen as a byproduct:

$$6\,CO_2 + 6\,H_2O \;\rightleftharpoons\; C_6H_{12}O_6 + 6\,O_2$$

In respiration, the opposite occurs: oxygen is removed from the water and carbon dioxide is released.

Now consider the equilibrium existing between carbon dioxide and carbonic acid into which carbon dioxide is partly transformed when dissolving in water:

$$CO_2 + H_2O \;\rightleftharpoons\; H_2CO_3$$

Adding extra carbon dioxide via respiration shifts the reaction to the right, resulting in the formation of more acid and in a lowering of the pH. Removal of carbon dioxide by photosynthesis in turn shifts the reaction to the left, decreasing the concentration of carbonic acid and thus increasing the pH.

To understand the buffering capacity of natural waters against such pH changes, one should recall that carbonic acid, in fact, partly dissociates into bicarbonate (HCO_3^-) and hydrogen ions (H^+). Bicarbonate ions also dissociate into carbonate ions ($CO_3^=$) and hydrogen ions (H^+). All of these reactions are reversible. In well-buffered waters that are rich in bicarbonate and carbonate ions, most of the hydrogen ions originating from the extra carbonic acid formed by intensive respiration will first be bound to the carbonate and then to the bicarbonate. Since the pH of water reflects the concentration of free hydrogen ions, the acidic effect induced by respiration can be neutralized to a large extent. In poorly buffered waters, however, hydrogen ions originating from the dissociation of extra carbonic acid due to respiration of the biota will rapidly exhaust the free carbonate and bicarbonate ions. This ensures that more hydrogen ions will remain dissociated, causing lower pH's.

The buffering capacity of freshwater is defined as bicarbonate and carbonate alkalinity and can be measured by titration with an acid and two indicators. The total alkalinity refers to the sum concentration of these two compounds expressed as mg/l of equivalent calcium carbonate. Total alkalinity is the most unambiguous measure of the buffering capacity of aquacultural waters. Water originating from areas with limestone is usually highly buffered and the alkalinity may exceed 100 to 200 mg/l. Conversely, water originating from sandy pine-woods areas may be poorly buffered with total alkalinity less than 25 mg/l.

Sometimes well-buffered waters are called "hard" waters, while poorly buffered waters are called "soft" waters. The concept of "total hardness" actually refers to the concentration of divalent cations in water (mainly calcium and magnesium) which are bound to the bicarbonate and carbonate anions. Hard water is usually richer in bicarbonates, carbonates, and other complex positive ions than soft water and therefore usually has better buffering capacity. Hardness is expressed in "degrees of hardness" as well as in mg/l of equivalent calcium carbonate. (A degree of hardness is equivalent to about 17.1 mg/l.)

In some cases, alkalinity may be high but hardness may be low in the absence of divalent cations. For example, marine waters have, in addition to a high carbonate alkalinity, a borate alkalinity, and they are quite well buffered against shifts in pH. In freshwaters, however, alkalinity and hardness are usually equivalent.

QUALITY OF INTAKE WATER

In nearly all production processes, water and raw materials are combined to obtain the desired product. Unfortunately, the reaction also produces wastes (see Figure 4.1). Since this also applies to aquaculture, both the quality of the incoming water as well as that of the outgoing water need to be taken into consideration.

With regard to the inflowing water, every aquacultural endeavor is critically dependent on a good water supply, in terms of both quantity and quality. During site selection this is the very first factor to be taken into account. Considering the many physical, chemical, and biological parameters that can influence water quality, as explained previously, intake waters for farming must be constantly monitored. Because water quality determines the well-being of the farmed organisms, it automatically influences growth rate, food conversion ratios, and sensitivity to diseases and toxicants. Some problems related to influent water quality not already discussed will be briefly discussed now.

FIGURE 4.1. Simplified scheme of a production process.

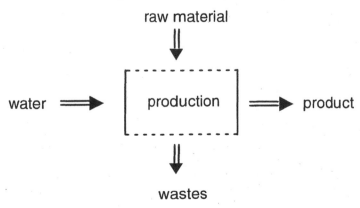

a. Turbidity. In some areas, particularly in estuaries, water may contain a high load of particulate material in suspension, up to 300 mg/l. This can seriously affect cultured biota by abrading or covering respiratory surfaces and smothering eggs and larvae. High levels of suspended solids can also greatly reduce light penetration, photosynthesis, and production of natural food items. Concentrations of suspended solids greater than 25 mg/l can adversely affect growth rate in some fishes and growth rate can be reduced up to 80% at turbidity levels above 100 mg/l.

b. Acidity. Domestic and industrial pollution adversely affect the pH of soils and waters in some parts of the world and may well become a serious water quality problem for aquaculturists. In addition, as waters acidify and pH declines, some metals such as aluminum in the soil and in aquatic sediments will become solubilized. When aluminum ions are solubilized, they react with water to form hydroxides that precipitate on the gills of fish and interfere with respiration. Most heavy metal ions that are liberated by acidification of water are toxic to aquatic biota, and their bioaccumulation in the flesh of farmed species may constitute a hazard to human health. Hard waters are less affected because they are not only well-buffered against changes in acidity, but their high concentration of calcium carbonate counteracts the toxic effects of metal ions in solution. This is illustrated in Table 4.11, which shows the increasing tolerance of some fish species to zinc as water hardness increases.

c. Eutrophication. The release of high levels of organic wastes and soluble nitrogen and phosphorus compounds from aquaculture systems can lead to very serious water quality problems in receiving waters. Aerobic microbial degradation of organic wastes can decrease the oxygen content in shallow confined areas to stressful and even lethal levels. Toxic levels of ammonia and nitrite may also accumulate. Algal blooms invariably develop in receiving waters where effluents are rich in nitrates and phosphates. These affect desirable biota in several ways, both in the natural system and downstream or adjacent aquaculture systems.

First, during the day, the pH of receiving waters may rise substantially because of the strong photosynthetic activity, which in turn will negatively influence the percentage of ammonia that is in

TABLE 4.11. Tolerance limits for zinc for coarse fish (Cyprinids) and salmonids in waters of increasing hardness.

Water hardness	Zinc mg/ l	
mg / l CaCO_3	Cyprinids	Salmonids
10	0.3	0.03
50	0.7	0.2
100	1.0	0.3
500	2.0	0.5

the toxic un-ionized form. Second, during the night, the oxygen content of the water may, become depleted because of respiration of the phytoplankton and decomposers in the system. Third, several microalgal species (including bluegreen algae) that thrive in enriched environments are known to be toxic either to desirable aquatic species or to mammals, including humans, that may ingest aquatic foods contaminated by these algae. Nontoxic algae may also generate soluble, "off-flavor" compounds that render cultured species unpalatable. Development of microalgal blooms has in practically all cases had a serious negative impact on aquacultural operations.

· *d. Pathogens.* Several kinds of pathogens (bacteria, viruses, parasites) can be present in intake waters if these originate from an area in which effluents of domestic wastes are discharged. The pathogens may cause diseases in the farmed species or they may accumulate in the organisms and thus constitute a threat to the consumer. In recent years there has been speculation that high intensity fish culture operations in Asia, where a variety of species such as ducks, swine, and humans have been in close association, have acted as "incubators" in which new and dangerous strains of human pathogens (influenza) have been formed.

e. Industrial toxicants. As previously mentioned, there are aquaculture operations that use the heated effluent from the cooling waters of power plants. In order to prevent fouling of the pipes, chlorine or hypochlorite is added at regular intervals to the heated effluent. Both compounds are toxic to aquatic life–hence their use to combat fouling organisms–and they can also affect species cultured in such waters. Brown trout can be killed by a concentration as low as 10 micrograms/l of Cl_2 which corresponds to 8 micrograms/l of hypochlorite.

Researchers have noted several even more insidious types of pollution that can ruin water quality in aquaculture systems. Pollution of natural waters by inorganic compounds such as heavy metals or by organic compounds such as pesticides may directly affect aquatic species. The direct toxicity of some of these compounds should not be underestimated for aquaculture species. Concentrations of some pesticides as low as 0.01 micrograms/l can induce chronic health problems. Such very low levels may well occur in shallow areas where pesticides are sprayed on agricultural crops nearby. These pesticides can reach the aquaculture facility through aerial drift or through both ground and surface water sources as they leach from the fields.

These toxins may also bioaccumulate and constitute a hazard to the consumer. In bioaccumulation, relatively low levels of a toxin dissolved in the water can accumulate in the tissues of aquatic organisms (e.g., zooplankton) that are in the food chain. Animals progressively higher on the food chain will integrate progressively higher concentrations of the toxin in their tissues. Bioaccumulation is especially potent with lipid-soluble pesticides and these are sometimes found in concentrations many thousands of times higher in fish tissues than in the fish's environment.

f. COD and BOD. COD refers to Chemical Oxygen Demand and BOD to Biochemical Oxygen Demand. The COD represents the total amount of oxygen necessary to oxidize organic materials in water. The BOD represents the total amount of oxygen necessary to oxidize those organic materials that can be used by microbial organisms present in the ecosystem. Both measurements are usually reported with a subscript denoting the number of days required for the process to be completed. COD is often calculated for effluents from factories, whereas BOD is most often used to measure the organic load generated by domestic wastes, including human wastes and food processing wastes. High levels of either can lower oxygen to lethal levels.

EFFLUENT QUALITY AND TREATMENT METHODS

Substantial volumes of water may be discharged from aquaculture units, especially flow-through systems. Such waters contain dissolved as well as particulate organic materials and are usually rich in nutrients such as phosphorus and nitrogen. Even when the concentrations of wastes in the effluents are low they can be significant because of the high water volumes involved. Therefore, wastes must be removed or treated to prevent environmental pollution. Waste loads in aquaculture effluents can be evaluated by conversion to "person equivalents" used for effluents discharged in urban areas. A "person equivalent" corresponds to 12 g nitrogen/day, 2.5 g phosphorus/day and 75 g oxygen/day (as BOD_7). It is easy to calculate the amount of waste in person equivalents from aquacultural units if the phosphorus content of the feed, the biomass of fish, and the food conversion ratio are known. There are, however, marked differences between the effluents discharged from an urban area that have passed through a water treatment plant and the effluents from an aquacultural unit. The ratio between carbon, nitrogen, and phosphorus varies much in the two types of effluents, as does the ratio of suspended solids to dissolved substances. An "average" effluent from a finfish aquaculture unit has the composition shown in Table 4.12.

The composition of the feed as well as the feeding technique influence the amount of wastes produced. For example, the estimated daily effluent phosphorus load per kg of fish produced is influenced by feed phosphorus content and feed conversion ratio as shown in Table 4.13. Note that as phosphorus content and feed conversion ratio increase, the amount of effluent phosphorus increases.

The impact on the environment of a commercial sized rearing unit annually is very substantial, as can be seen from Table 4.14. Research has shown that the largest part of the waste is associated with suspended solids and most can be removed from the water if satisfactory methods are used for separation of suspended solids.

TABLE 4.12. Average amount of wastes in an effluent from a culture unit in mg/ l.

Organic material	
BOD	5-20
COD	10-185
Suspended material	
SS	5-50
Phosphorus	
Total P	0.05-0.20
Nitrogen	
Total N	0.3-4.0

Some 70% of the COD, about 50% of total nitrogen, and about 50% to 70% of the total phosphorus can be eliminated by such "primary" treatment.

There are many methods for removing suspended solids. These include: straining, filtration (trickling), sedimentation, flotation, and whirlpool separation. The costs involved in using such mechanical systems to separate suspended solids from effluent waters can be significant. The annual costs of removing suspended solids from the effluent of a unit rearing one million Atlantic salmon smolts with a water supply of 20 m^3/minute varies from $60,000 to $800,000 (U.S.).

Note that effluents from static culture systems may contain very high levels of suspended solids but are discharged infrequently. Thus, regulatory agencies classify them differently from flow-through systems and rarely require that waste treatment systems be installed.

TABLE 4.13. Effluent phosphorus load as a function of feed conversion ratio and feed phosphorus content. Units are g of effluent phosphorus per kg of fish produced.

Feed conversion ratio	Feed phosphorus content		
	0.75%	1.25%	1.75%
1.0	6 g	10 g	16 g
1.5	8 g	14 g	22 g
2.0	10 g	18 g	28 g
2.5	12 g	22 g	34 g

TABLE 4.14. Pollutants from a rearing unit producing 500,000 Atlantic salmon smolts annually.

Effluent pollutant	kg/year
Organic material (COD)	75,000
Total phosphorus	600
Total nitrogen	3,800

Chapter 5

Energy Use in Aquaculture Production

Natural production in almost all ecosystems is based on solar energy. Solar energy is converted to chemical energy by photosynthesis (primary production) and then this energy is transferred by consumers to successive trophic levels in interlinked food chains (or food webs). The only known ecosystems not based on solar energy are the recently discovered specialized ones associated with deep-sea thermal vents. In all ecosystems the transfer of energy from one link to another in food chains always involves some losses. Ecologists estimate that there is an average loss of 90% of energy at each level of the food chain. For this reason, shorter food chains are more efficient than longer ones and most natural systems do not have chains longer than five or six links. These inefficiencies are also partially responsible for the fact that offshore sea areas where the food chains are long (five links) produce less fisheries protein per unit area than coastal areas where the food chains are shorter (three links). In addition, the concentration of plant nutrients in offshore areas is usually less than in coastal ones.

In contrast to wild fishes, animals in aquatic farms are fed with manufactured dry feed, semimoist feed, wet feed (minced fish), and/or vegetable matter. The raw materials of these feeds consist of standard animal feed ingredients such as fish, fish meal, grain and grain products and byproducts of meat processing industries. Formulated feeds are frequently enriched with minerals and vitamins. Energy must be provided in the raw feedstuffs, the food manufacturing process, and in the farming process itself.

To summarize, the energy utilized in aquaculture systems is derived from four sources:

1. Solar energy bound in natural production.
2. Energy used for the preparation/manufacturing of feeds, equipment, and supplies used in the aquaculture process.
3. Energy used during the aquaculture process.
4. Energy bound in the organic materials provided to the cultured species.

In aquaculture systems, then, another step is added to the food chain when processed wild fish (fish meal or minced fish) is used as a food source to raise another species of fish. Some ecologists contend that extending the food chain by using fish meal to produce farmed fish is not the most efficient way to utilize resources from the marine environment. However, the main alternative use for these products is in feeds for terrestrial animals and these generally have lower efficiencies than aquatic ones. The energy of digested food is usually better utilized in fish than in terrestrial animals. It is apportioned as follows: 28% for growth, 43% for life maintenance, and 27% for excretory functions.

The balance of energy input (farming) and output (harvest) can be calculated by determining the energetic needs of the different processes and the energy content of the product itself. Table 5.1 shows the amount of energy involved in producing 1 kg of rainbow trout. By applying the figures from Table 5.1 to a net cage operation, it is possible to calculate the energy balance for a 50 metric ton production unit (Table 5.2). The figures provided in Table 5.2 are based on the following assumptions:

1. The average weight of "seed" fingerlings is 80 g.
2. The individual fish grow from 80 g to 2,800 g during a period of 360 to 420 days.
3. The feed conversion ratio for juvenile fish is 1.25 for dry feed and 2.2 for semimoist feed, and the growth factor is 10.
4. The juvenile mortality rate is 1.0%.
5. The feed conversion ratio for preadult and adult fish is 1.5 for dry feed, and 2.5 for semimoist feed, and the growth rate is 3.5.
6. The adult mortality rate is 1.0%.
7. The average loss at slaughter is 36%.

TABLE 5.1. Amount of energy involved in production of 1 kg of rainbow trout. Source: Leppäkoski.

Production Factors and Components		Energy (in 10^6 joules)
Dry feed	Manufacturing	24.1
	Energy content	21.0
	Total	45.1
Semimoist feed	Manufacturing	26.0
	Energy content	11.6
	Total	37.6
Seed	Total	86.4
Production unit	Labor	0.18
	Equipment	27.0
	Total	27.18
Product	Gross weight	6.5
	Net weight (fillet)	9.2
	Gonads	5.4

From the data in Table 5.2, it follows that aquaculture is not a very energy-efficient endeavor in terms of input versus output of energy. Indeed one must supply 11 to 16 times more energy than one realizes. It is, therefore, interesting to compare the energy expenditures for several forms of aquaculture with the energy used in plant and animal husbandry (Table 5.3). The figures in Table 5.3 demonstrate that the energy spent for various types of fish production is usually lower than that for production of animals on land. The energy expenditure for chicken production in the U.S.A. is comparable with the highest fish production values shown, while the production of beef requires four times as much energy. There are, however, high energy aquaculture systems as well. For example, energy expenditure figures for prawn and shrimp production (not shown) in some aquaculture systems are on the same order as those for beef production. Moreover, traditional fisheries can have very high energy expenditures. For example, long distance trawling fleets require substantial amounts of energy to hunt, capture, and transport fish and fish products.

TABLE 5.2. Energy budget in gigajoules (GJ) for the production of 50 metric tons of rainbow trout in net cages. Source: Leppäkoski.

	Energy Input or Output			
	Dry feed		Semimoist feed	
	GJ	%	GJ	%
Feed	3,978	94.5	5,719	96.2
Seed	155	3.7	155	2.5
Equipment	68	1.6	68	1.1
Labor	9	0.2	9	0.2
Total Input	4,210	100	5,951	100
Edible fish	369		369	
Output Efficiency	0.088 (8.8%)		0.063 (6.3%)	

TABLE 5.3. Energy expenditure for production of various foods expressed in kcal or 10^3 joule per gram unprocessed protein.

Food item	kcal per g	10^3 Joule per g
Wheat (USA)	13.7	57.3
Rice (USA)	40	167.4
Chicken (USA)	149	623.6
Beef (USA)	685	2,866.7
Catfish (Thailand, USA)	83-165	347.3-690.5
Carp (Indonesia, Israel)	6-53	25.1-221.8
Tilapia (Africa, Israel)	4-65	167.0-272.0
Salmon (Japan, USA, UK)	30-159	125.6-665.4

Chapter 6

Important Components of Aquaculture

INTRODUCTION

In the previous sections, we have focused mainly on the major prerequisite for successful aquaculture, water itself. Several other categories of factors, biological, technical, and socio-economical, must be considered and addressed, as well, in order to produce an acceptable and profitable product. The interchange between various factors, energy flows, and feedback mechanisms are outlined in Figure 6.1. Some of these factors are addressed in this section and others in following sections.

BIOLOGICAL FACTORS

The basic biology of a species, plant, animal, or microbe must be well understood before it can be successfully cultivated. While it may seem that more than enough data are available to successfully cultivate a species, initial cultivation efforts invariably and quickly show gaps in information that must be filled if success is to be achieved.

Confining fish and other aquatic organisms at high densities is usually necessary for economically viable aquaculture. However, such densities introduce problems that do not exist or are of minor importance under natural conditions. These include strain selection, breeding, nutrition, and diseases.

Strain Selection and Breeding

One selects species and strains in order to obtain high growth rates, optimal feed conversion efficiencies, late maturation, relatively nonaggressive behavior, tolerance of crowding and unfavorable water quality conditions, and high disease resistance. This has historically been accomplished by breeding and crossbreeding. Newer methods involve the production of unisex populations by hormone treatments of juveniles in combination with crossbreeding, the production of triploid specimens, hybridization, and, most recently, genetic engineering.

A selective breeding program should consider the needs of both the industry and the consumer. From the farmer's point of view, high growth rate, optimal food conversion efficiency, late maturity, good disease resistance, etc., may be the most important considerations. However, the consumer may be more concerned with meat quality with regards to meat yield, fat content, color, taste, body shape, dressing percentage, etc. The overall objectives for all breeding programs, then, are to (1) increase productivity and product quality, and (2) develop animals that are better adapted to captivity.

Several examples of selective breeding illustrate the progress and prospects within this area. An annual increase of 3 to 5% in yield has been achieved with some salmonids. Maturation has been delayed one to two years in some strains of rainbow trout. The development of unisex populations of tilapias has increased production significantly. Improved disease resistance in some molluskan species has been achieved.

Nutrition

The nutritional requirements of fish and shellfish are essential data for their successful cultivation. Feed usually accounts for 40 to 50% of the production costs in semi-intensive and intensive aquaculture systems. This is because many farmed species are carnivorous and require much more protein, especially animal protein, than do most domesticated terrestrial animals. In contrast, however, aquatic animals require less energy for protein synthesis. Fish use much less energy in terms of protein gain per calorie of metabolizable energy (ME) consumed (Table 6.1).

There are several reasons why fishes and shellfishes have a lower dietary energy requirement than domestic land animals. First, they are poikilothermic: unlike homeothermic animals, they do not have to maintain a constant body temperature. Second, they need less

FIGURE 6.1. A scheme of important biological, technical, and socio-economical factors influencing the development of aquaculture. Modified after Ackefors and Roséll (1979).

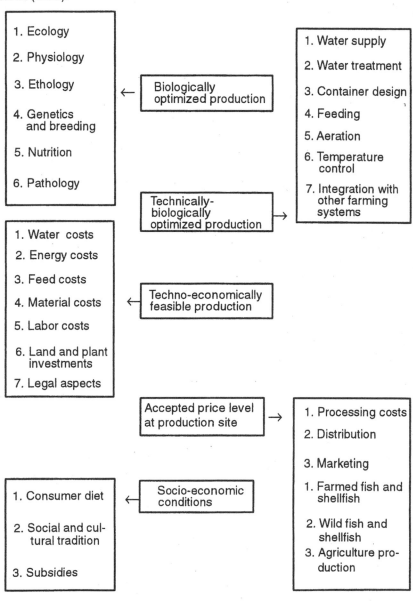

energy to maintain position and to move in water than animals do on land. Finally, they excrete most nitrogenous wastes as ammonia rather than urea and uric acid, the main excretory products for terrestrial animals; therefore, energy loss in protein metabolism is much lower. This phenomenon is also illustrated by comparing the feed conversion efficiencies for protein and energy in the production of edible flesh of salmon, swine, and chickens (broilers) (Table 6.2.).

All animals require an energy source, essential amino acids, essential fatty acids, fiber, specific vitamins, and specific minerals to thrive. The protein component of fish feed is normally very high–35 to 50% of dry feeds versus 10 to 20% in feeds for terrestrial food animals. Since animal protein is more expensive than vegetable protein, nutritionists strive to reduce the animal protein in feeds

TABLE 6.1. Comparison between the protein requirements and the protein gain per calorie of metabolizable energy consumed for several terrestrial and aquatic food animals.

Animal	Dietary protein (%)	Dietary ME (kcal/g)	Protein gain per Mcal (ME) consumed (g)
Channel catfish	35	2.7	47
Chicken/broiler	18	2.8	23
Swine	16	3.3	9
Beef cattle	11	2.6	6

TABLE 6.2. Food conversion efficiencies for protein and energy in the production of edible flesh of salmon, swine, and broiler chickens. Source: Austreng.

Food Conversion Efficiency	Salmon (1-5 kg)	Swine (1-100 kg)	Chicken (1-2 kg)
Gross energy %	27	16	11
Metabolizable energy %	39	20	16
Protein %	33	16	22

through use of vegetable protein sources including soybean meal, wheat middlings, and rice bran. Least-cost, formulated feeds are prepared using linear programming methods in feed mills. The source of nutrients for each cultured species must be chosen with care. In most fishes, ten amino acids are essential and cannot be omitted from the diet: arginine, histidine, isoleucine, leucine, lysine, methionine, phenylalanine, threonine, tryptophan, and valine. Furthermore, the relative proportions of these required amino acids in the diet must be correct in order to meet dietary needs for each species cultured. Sufficient sources of energy other than protein must also be included in feeds in order to spare as much protein as possible; otherwise, expensive protein is utilized as an energy source. A diet well-balanced with carbohydrates and lipids will spare protein, thus reducing costs and often giving rise to enhanced growth rates.

Aquatic species require lipids for supplying energy, for synthesizing cellular structure, and for maintaining the integrity of biomembranes. In addition, dietary lipids provide the vehicle for absorption of fat-soluble vitamins and also provide other compounds such as sterols, essential in the diets of crustaceans. Essential fatty acid (EFA)[1] requirements vary widely in the animal kingdom. In terrestrial homeotherms such as mammals, the so-called n-6 and n-9 series of fatty acids dominate. In aquatic poikilotherms, temperatures lower than 20°C favor metabolism based on the more unsaturated n-3 series of fatty acids. Although most fish can convert the n-6 to the n-3 series, they cannot cope with large proportions of n-6 lipids. The coldwater fishes have limited ability to synthesize fatty acids of the n-3 and n-6 series. This is important because the n-3 fatty acids are essential for maintenance of good health and promotion of rapid growth in a number of species.

Carbohydrate is not an important component of the natural diets of most aquatic animals, even though all species have a sufficient complement of digestive enzymes to cope with some starches and disaccharides. Utilization of carbohydrates, however, is highly spe-

---

1. Fatty acids are described by the formula x:y n − z where x is the number of carbon atoms, y is the number of double bonds, and n (formerly omega) − z is the position of the first double bond counting from the methyl end of the fatty acid.

cies specific. For example, sucrose and lactose can be readily assimilated by rainbow trout but the species cannot assimilate potato starch and dextrin well. Some vegetarian species such as grass carp can use large quantities of carbohydrates, but these have an unusual diet.

Diseases

Fish diseases and parasites often cause great losses in aquaculture operations. Diseases may be transferred between wild and cultivated fishes. Cultivating fishes at high densities facilitates the spreading of pathogens. Disease agents include viruses, bacteria, fungi, and protozoan and other parasites. Many agents that cause diseases under culture conditions are commonly present in the aquatic environment and become a problem only under stressful conditions. Other agents are not ubiquitous but are highly virulent and will kill a high percentage of otherwise healthy fish if introduced to a culture facility.

Information about fish viruses is limited. Viruses are often carried by fish with no clinical symptoms, and there are seldom any treatments available when an epidemic occurs. Drying and disinfecting production units is sometimes necessary to eliminate viral pathogens. Infectious hematopoietic necrosis (IHN), viral hemorrhagic septicemia (VHS), and infectious pancreatic necrosis (IPN) are very real threats to aquaculture operations.

Bacterial fish diseases have become very serious problems in aquaculture systems. However, unlike viruses, these diseases can be treated with antibiotics and chemicals in baths or feeds. In addition, some can be controlled by vaccination. Vibriosis, furunculosis, enteric redmouth disease, and bacterial kidney disease are examples of important bacterial fish diseases.

Fungus infections and protozoan parasites can also be deadly to cultured fishes. Most of these, such as the fungus *Saprolegnia* and the protozoans *Ichthyophthirius* (Ich) and *Trichodina* respond well to chemical treatments. However, others, like the sporozoan *Henneguya*, are intracellular parasites that are difficult to treat chemically. The critical step in treating bacterial and parasitic fish diseases is to get early diagnosis and identification of the pathogen so that the correct drug or chemical treatment can be used.

Shellfishes are also affected by parasite and disease problems. The parasitic copepod *Mytilicola intestinalis* causes great losses in mussel culture, as does the protozoan *Bonamia* in oyster culture. Bacteria of the genus *Gaffkemia* cause major losses in lobster rearing and holding systems. The crayfish fungus plague, *Aphanomyces astaci*, has decimated European freshwater crayfish populations for over a century.

Aquatic species are especially susceptible to attacks by disease and parasites when they are under stress. Proper husbandry of the animals is the best way to prevent such problems. It is also important to prevent transfer of parasites and diseases between farms and regions. This requires regulation of such movements, especially between countries, and should include strict quarantine measures.

The aquaculturist must realize that his or her crop can also create human health problems. Some parasitic helminths can infect humans when the host fish is eaten raw. Bivalve mollusks, especially mussels, clams, and oysters, may concentrate toxic compounds from various species of algae that they eat to the point that their consumption is potentially life-threatening. Diseases such as cholera and hepatitis may be contracted by humans eating improperly prepared fisheries products.

TECHNICAL-BIOLOGICAL FACTORS

The importance of a good water source cannot be overemphasized. Large quantities of high quality water are needed for flow-through systems. When water quantity is limited, the water may be treated and then reused within closed, recirculating systems with a very low turnover rate. In either case, the oxygen concentration must be kept fairly high, 5 to 9 mg/l, and the concentration of metabolic products must be kept as low as possible. Un-ionized ammonia and nitrite are highly toxic substances and cannot be permitted to accumulate.

Optimal water quality is also a function of the container design. Very few basins, tanks, or troughs have a shape that ensures uniform mixing even though there is a continuous supply of aerated water. Fish and crustaceans tend to aggregate in certain parts of a container where conditions are more favorable. Poorly designed containers have dead zones, especially in corners, where settleable solids will accumulate and decompose. These areas will have reduced the levels of DO and increased levels of CO_2, H_2S, and NH_3. The diameter of the inlet and outlet and the relationship between height and width of the container are essential factors to be considered when containers are designed.

Optimal feeding techniques are essential for promoting good growth and preventing the fouling of water with wasted feed. Feeding behavior and the interaction between individual animals are influenced markedly by stocking density. They are of great significance when the feeding strategy is planned. This is especially true for young fish; swim-up fry (youngsters just beginning to feed) may require feeding 12 or more times per day. Uniform distribution of the feed is also important to prevent nonuniform growth of young fish that can lead to cannibalism in many species. Later, as the fish get older, feeding frequency is gradually reduced to one or two periods per day for most species. For optimum growth, feed conversion, and well-being, the fish must be fed at optimal intervals and with a nutritionally adequate, properly sized feed.

The integration of aquaculture and conventional terrestrial agriculture can maximize profitability under the proper conditions. This is especially effective when low-trophic-level aquatic species are cultivated together in polyculture. This is seen in Asia where livestock such as dairy cattle, chickens, ducks, and pigs are raised on the same farms along with fish and crustaceans. Wasted feed is eaten directly by the fish and crustaceans, and the animal manure fertilizes pond waters, generating benthic and planktonic food organisms. Asian paddy rice fields are also used to cultivate hardy species capable of growing in shallow, still waters. Yields are low but costs are minimal.

Integration of agriculture and aquaculture is also practiced in industrialized countries where wastes or cultivated crops are used directly as fish feeds or in their manufacture. In the U.S., some trout farms are made profitable as a consequence of the simultaneous cultivation of corn and other crops used in the manufacture of fish feeds. Crayfish are cultivated in rice fields during cool months in southern states where they thrive on decomposing rice stubble. Some industrial wastes can even be used as fish feeds or as substrates for producing single cell yeast "protein," which then can be incorporated in fish feeds.

Integrated farming and polyculture are more ecologically sound methods of farming than monoculture because matter and energy are used more efficiently. Such systems tend to be more profitable than monoculture, especially in those parts of the world with high energy prices and low labor costs.

LEGAL-ENVIRONMENTAL FACTORS

The legal aspects of aquaculture are sometimes complicated. Many acts and ordinances that affect aquaculture were instituted before aquaculture became an established industry. In some localities certain species of fish may be legally classified as gamefish and may not be sold commercially even when raised under cultured conditions. In the U.S., strict regulations normally prohibit new construction, including fish ponds, in areas that are designated as wetlands. Even in upland sites in the U.S. it generally requires six to 18 months to obtain the necessary permits and licenses for construction and operation of a new aquaculture facility.

An aquaculturist must have access to water and land. National legislation may be complicated and make it difficult to obtain the right to use land even if water is available, or vice versa. An especially vexing problem is locating aquaculture ventures in areas where there are several different users of the site. Lakes and coastal areas are used for recreation, harbors, shipping, industry, housing, and military purposes. Established legislation often favors such interests over those of aquaculture. The attitudes of the national, regional, and local regulatory bodies and the general public are decisive in determining the resolution of such conflicts.

The environmental aspects of aquaculture have become important considerations in its development. Rising consciousness for conservation of natural resources has heightened interest in legislation affecting aquaculturists. Pollution from the discharge of waste waters from communities, industries, forestry, and agriculture has precipitated new legislation to correct such problems in many countries. Waste treatment facilities have been built to reduce the discharge of harmful compounds. Aquaculture does not differ from other water-consuming businesses. Discharge of organic wastes and nutrients from aquaculture units is a real problem that cannot be ignored. Such materials may change the ecological balance in receiving waters. Oxygen depletion in bottom waters, toxic algal blooms, and changes in the natural species composition are several potential negative affects of aquaculture that must be avoided as much as possible.

Eutrophication of coastal waters has become a serious problem in certain parts of the world, although aquaculture has usually played a comparatively small part in that problem. Nutrients discharged by conventional agriculture, industry, forestry, and the atmosphere usually have much greater impacts on the aquatic environment than aquacultural endeavors. However, it is much easier for authorities to calculate the discharge from aquacultural operations than from these other activities, which are more diffuse in character. Legislation designed to preserve the aquatic environment has, therefore, been applied so rigorously in some countries that aquaculture development has been seriously inhibited. The aquaculture industry has, nevertheless grown–in fact, in some countries, too rapidly and without regulation, creating real environmental problems.

TECHNICAL-ECONOMIC FACTORS

This is certainly one of the most intricate areas for an entrepreneur to address. The costs of machinery, equipment, energy, feed and seed, interest rates, and so forth must be balanced against the selling price of the product. Variation in interest rates, market demand, and climate is beyond the control of the farmer. Information on writing a business plan and budget for various aquacultural enterprises is readily available (usually from a local or state agricultural extension office), but the costs are site-specific. It is beyond the scope of this book to go into details for such budgets. However, the entrepreneur should use such budgets as guides in making economic plans. They serve as a checklist to ensure that factors critical to financial success are not overlooked.

Production costs must be considerably lower than retail market prices for an aquaculture venture to be economically viable. There are substantial costs involved in processing (filleting, smoking, freezing, etc.), distributing, and marketing the aquatic products. Farmers often have difficulties accepting this fact. Comparatively low profits have often tempted producers to process, distribute, and even market their own products. These can be viable options if there are suitable local markets. However, there are frequently serious disadvantages to these activities both for the producer and for the industry itself. With the exception of local markets, it is impossible for an expanding aquaculture industry to survive if the processing, distributing and marketing are not handled by professional people. This is especially true for marketing aimed at export to other countries. Such business is very complicated, involving customs documentation, health certificates, import regulations, and special demands of a foreign market.

SOCIO-ECONOMIC FACTORS

Various socio-economic conditions will greatly influence the accepted price of an aquaculture product. Local fish and shellfish consumption customs are likely to be the key factors in determining market demand. Social and cultural traditions influence not only the choice between fish and red meat but also the kind of fish people prefer. There is a strong tendency to prefer a popular, expensive species rather than a cheaper, unknown species.

Competition from wild fishery products can be a problem for the aquaculture industry. However, in some countries, consumption actually increased when cultured fish became available because aquaculture stabilized supplies. The subsidizing of food costs can have a positive influence on food consumption patterns. At present, red meat consumption is decreasing in many countries due to higher prices, even though people generally prefer to eat red meat. An interesting trend, however, has developed in many industrialized countries: the health benefits of consuming fish are shaping consumer tastes. Fish are low in fat and have a high level of unsaturated fatty acids, both of which are desirable for human health. For this reason even high fat fish species like eel, salmon, and mackerel are good for the consumer. In addition, fish flesh offers the consumer valuable micronutrients and amino acid profiles (the correct ones in the correct proportions).

Chapter 7

Factors Involved in the Selection of Species Suitable for Aquaculture

INTRODUCTION

Species or strain selection is a most important decision for the aquacultural entrepreneur. First and foremost, one must consider the past history of the species in other cultural situations. A knowledge of that species' biological needs must be considered in relation to site-specific climatological considerations and quality and quantity of water. These factors are addressed in this chapter.

TEMPERATURE

It is generally impractical to control ambient temperatures in aquaculture systems. Therefore, local climate dictates operational temperatures. The mean, maximum, and minimum temperatures in the water are controlling factors when a choice of species is made. Even though many species tolerate a wide spectrum of water temperatures, the optimum temperature range for good growth rate, low feed conversion ratios, reproduction, and disease resistance must be considered. In temperate climates, thermal minima are often serious constraints to aquaculture. Overwintering of fish may be a critical problem. Ice conditions can be an obstacle if not anticipated in advance. Some species such as oysters cannot endure freezing conditions. Supercooled water is a very serious problem at high latitudes or altitudes; lethal temperatures for coldwater fishes such as salmonids are about −0.5°C.

Thermal maxima must also be considered. As with minima, these are influenced by the acclimatization temperature and can fluctuate as much as 5°C for the same species. Some representative thermal maxima include: salmon, 22 to 24°C; rainbow trout, 25 to 26°C; brown trout, 25 to 30°C; and carp, 40 to 41°C (Table 4.5). High, life-threatening temperatures may be experienced at high latitudes during mid-summer when solar exposure is almost constant.

SALINITY

Many cultured species are euryhaline, that is, they thrive over a wide range of salinities. In Thailand, groupers can grow well in estuarine areas where there are wide daily fluctuations in salinity. Some tilapia species may be cultivated easily in high-salinity waters although they are originally freshwater species. Eels can accept brackish to marine waters for the grow-out period even though the females live naturally in freshwater during this part of the life cycle. The channel catfish grows well in salinities up to 8 ‰ although it is clearly a freshwater species.

Salinity tolerance in some species is a function of life cycle stage. The freshwater prawns, *Macrobrachium* spp., and mullets, *Mugil* spp., are catadromous–they spawn in brackish estuarine or marine waters, undergo larval development there, and move into freshwater where they mature. Other important species including striped basses, salmon, brook trout, and Arctic char are anadromous–they spawn in freshwater but mature in seawater. Rainbow trout, a landlocked anadromous species, can be cultivated in full-strength seawater. However, they become very sensitive to saline waters when they mature. High mortality rates are observed in males in sea pens. It is therefore important to know the salinity tolerance limits for each species in its various developmental stages.

DISSOLVED GASES

Dissolved gases have been discussed previously. In this section, however, we reemphasize the fact that aquatic species require oxygen-rich waters for optimum growth and survival (Illustration 7.1).

There are species like the carps that tolerate much lower oxygen levels than species such as salmonids; but important economic considerations such as growth rate, long-term stress, food conversion, disease resistance, etc., are adversely affected when any species is continually exposed to low oxygen levels, even though such low levels are not life threatening. A few hardy species have accessory respiratory organs that allow them to use atmospheric oxygen. These "air-breathing" fishes can function well even in waters with serious oxygen depletion problems. Examples include the gouramis and snakeheads of Asia, walking catfishes of Africa, and the bowfin in North America.

ILLUSTRATION 7.1. Paddle wheel aerator is used for emergency aeration in channel catfish ponds and for preventing thermal stratification. Alfalfa valve in foreground breaks well water into small droplets to oxygenate water. Both kinds of equipment are mandatory to maintain dissolved oxygen levels in intensive pond culture of fishes and crustaceans. Credit: Jay Huner.

REPRODUCTION

Some species do not spawn in captivity. Their culture depends on capture of wild seed. Other species do reproduce in captivity. These may be divided into four categories:

1. Species that are spawned artificially with milt (fish sperm) and eggs that are stripped from parent fish without the use of hormones. An important example is the salmonid fishes.

2. Species that are spawned by first injecting hormones into mature parent fish and then either stripping milt and eggs or permitting the parents to spawn freely. Examples include the striped basses and the Chinese carps (but not the crucian carp, goldfish, or common carp which do not require hormone injection).

3. Species that are spawned by manipulating environmental cues such as temperature and photoperiod. Annual cycles may be compressed several times in a year to induce out-of-season spawning. Examples include turbot, European crayfish, seabream, abalone, oysters, penaeid shrimps, and red drum. Some species such as the crayfish and oysters will spawn readily in a normal seasonal sequence. Others will not spawn without manipulation. In addition to environmental cue manipulation, penaeid shrimps are often induced to spawn by removing (ablating) a single eyestalk. This reduces the levels of circulating ovarian inhibiting hormone(s).

4. Species that spawn readily without any manipulation other than to bring separated parent fish together or to provide spawning surfaces or containers. Examples include tilapias, sunfishes, ictalurid catfishes, common carp, freshwater crayfishes, freshwater prawns and many bivalve mollusks.

The larvae of many freshwater and anadromous species like the salmonids are easy to rear. Such species have comparatively big larvae that can swallow roe, minced liver, or formulated feeds. Conversely, most marine species such as penaeid shrimps, cod, flat fishes, sea basses, etc., produce small eggs, and the newly hatched larvae are generally very small. This complicates their cultivation, as they require living foods such as microalgae, brine shrimp, rotifers, and other zooplankton. Micro-encapsulated feeds are now being developed to replace such living feeds, but truly effective artificial feeds for many of these larvae have yet to be mass-produced. A common cultural technique remains the seeding of plankton-rich, heavily fertilized ponds with newly hatched fish larvae that cannot otherwise be provided with the necessary living feeds. Examples include pike-perch, black basses, striped basses, graylings, and whitefishes.

GROWTH RATES

A rapid growth rate that permits harvest in the shortest time is highly desirable in any species. It is economically most beneficial if a species can be marketed while it is still in the exponential growth phase of its life cycle.

Once obtained, fertilized eggs may be incubated in containers such as basins, trays, jars, and tanks. This is particularly useful with species such as striped bass, grass carp, and oysters that provide little or no care to the fertilized eggs. Some other species conveniently incubate their eggs through part or all of the larval period. Examples include freshwater prawns and crayfishes, homarid lobsters, catfishes, and tilapias. For some of these, such as catfish, there is better survival with artificial incubation and it is therefore preferred. In others, such as crayfishes, natural incubation on the mother is used even when done in hatchery trays.

Growth rates are temperature-dependent within a species' physiological tolerance limits. Aquaculturists in temperate climates are using intensive cultivation systems to reduce three- and four-year-long production cycles for salmonid species. In such cases, they culture small fish in energy-conserving systems and then grow them to market size in pens and raceways during short spring-summer-autumn growing periods. Most aquatic species have rapid growth rates during periods when environmental conditions are optimal; however, such periods are very short at high latitudes or altitudes. Thus, warm water species appear to have more rapid growth rates because they reach usable sizes much more rapidly than cold and coolwater species. For example, the red swamp crayfish from the southern U.S.A. reaches market size of 9 to 10 cm in three to five months, while signal and noble crayfishes require two to four years to reach such sizes in Scandinavia. Cultivating rainbow trout smolts intensively in "warmed" hatcheries (10°C) in Finland has permitted production of 1 kg food fish in outdoor cages and raceways in 18 to 24 months, halving normal production time.

NUTRITIONAL REQUIREMENTS

Detritivorous and omnivorous species usually have lower protein requirements than carnivorous species and can use plant protein sources more effectively. Herbivorous species such as grass carp and wuchang fish and detritivores like freshwater crayfishes are far more economical to raise than carnivorous salmonids and catfishes because protein requirements are lower, and less expensive plant and microbial protein can be used as feeds or in feeds. However, it must be emphasized that larval and juvenile phases of vegetarian and detrital fish and crustacean species require high quality animal protein until they complete early development.

Protein requirements of cultivated aquatic species range from lows of around 25% to highs of around 55% of the total feed ingredients. This figure often varies as a function of the life cycle stage, being greatest for rapidly growing larval and juvenile stages.

An especially costly aspect of rearing of some aquatic species is their dependence during these early stages–especially the larval stage(s)–on living feeds, both plant and animal. The plants, single-celled algae, may be fed directly to species that will consume phytoplankton, or may be fed indirectly to others by using small planktonic animals, such as rotifers and brine shrimp nauplii as intermediates. Much effort is being devoted to the development of suitable manufactured feeds (micro-encapsulated feeds) to replace these living feeds in larval rearing systems.

The absolute amount of feed required for culture of larval and juvenile stages is relatively small because of their small size; this helps to limit the costs of feed, and thus, overall production costs. However, with some species such as pike-perch, sea basses, and whitefishes, so much of these living feeds are required that the only economically viable way to rear seed fish is to stock them in open ponds where natural foods are nurtured by fertilization. Similarly, in the cases of many bivalve mollusks such as mussels and oysters, motile species such as penaeid shrimps, and some finfishes such as mullets and milkfish, it is more profitable to let nature produce and nurture the seed and to then collect them for on-growing.

The advantage of cultivating filter-feeding molluskan species in open waters, where it is not necessary to supply feed, is clear. Equally favorable from the feeding standpoint is the release of carnivorous fishes into open waters after cultivation through the larval/juvenile stages. Typical recovery rates of between 1 and 5% of the released fish may or may not be economically viable, depending on the value of the final product. In the case of salmonid fishes it is economically marginal.

GENETICS

Selective breeding permits the development of strains with favorable cultivation traits, such as delayed maturation in rainbow trout, rapid growth in carps, attractive coloration in tilapias and koi carp, etc. Genetic engineering through the actual incorporation of growth hormone genes from another species into fish eggs is an experimental reality around the world. If growth performance meets projections and the treated fish can transmit the genes to the next generation, there should be a revolution in aquaculture production. There is, however, considerable opposition to the idea of growing genetically engineered fishes in open ponds where they might escape into the wild. Thus, trans-specific genetic manipulation is still largely a matter of discussion and experimentation. Other types of genetic manipulation such as thermal shock of the fertilized egg and gynogenesis (the egg is stimulated to develop, but only maternal chromosomes become part of the zygote) can be used to produce polyploid (organisms with three or more sets of chromosomes) fishes that are sterile and grow rapidly. These two techniques are already widely employed by producers of grass carp and rainbow trout.

MARKETING

The market potential for any aquaculture crop must be checked and specific buyers targeted before a cost-and-return analysis can be made. Some species have tremendous market potential and high prices. However, the cost of farming them may make it impossible to make a suitable profit.

In cases where market prices shift widely, it may be advisable to investigate vertical integration either individually or through a cooperative of producers. This would require development of hatcheries, grow-out facilities, processing units, and marketing channels. Where feed is an important cost factor, a decision will have to be made as to whether or not a feed processing plant must be incorporated into the unit. All possible marketing outlets should be investigated, including seed, stocking for put-and-take fishing, and food fish (Illustrations 7.2 and 7.3).

Long-term planning and flexibility are essential to the success of *any* business venture. Aquaculture ventures are no exceptions. Unrealistic expectations for market prices and demands will doom any otherwise well-planned aquaculture venture to failure. There have been many examples of such failures in the past two decades.

SPECIES CULTIVATED FOR OTHER PURPOSES

Aquaculture is not limited to the production of food species. Crustacean, molluskan, and fish seed are frequently grown for fisheries management purposes. Crustaceans and small fishes are often cultivated as bait for sport and/or commercial fisheries. Ornamental species constitute one of the highest-value aquaculture industries (Illustration 7.4). Some species are cultivated solely to serve as living food for pet fishes. Others are raised for use in laboratory and educational institutions. Fishes such as grass carp and mosquitofish are cultivated and sold to control noxious aquatic weeds and disease-vectoring mosquitoes. Oysters and freshwater mussels are grown to produce cultured pearls and crocodilians are cultured for their highly valued skins. Algae are being reared for extraction of chemicals, water treatment, and biogas production. The potential or already practicing aquaculturist should be very flexible in developing aquacultural endeavors.

ILLUSTRATION 7.2. Transporting live fish to market in barges with live wells in southern China. Credit: Jay Huner.

ILLUSTRATION 7.3. Transferring live channel catfish from holding pen after pond has been seined. Aerated water to right of pen maintains dissolved oxygen levels for penned fish. Credit: J. W. Avault, Jr., Louisiana Agricultural Experiment Station.

ILLUSTRATION 7.4. Goldfish, *Carassius Auratus*, are cultivated as live ornamental fish feeds, as ornamental fishes themselves, as food fishes, and as live fish baits. Credit: J. W. Avault, Jr., Louisiana Agricultural Experiment Station.

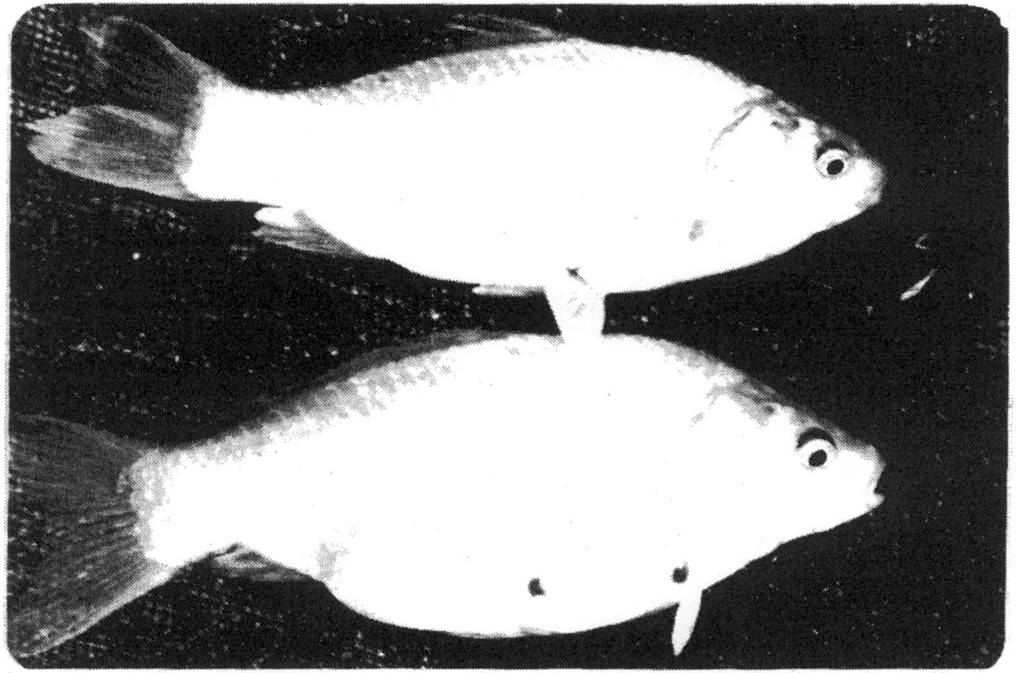

Chapter 8

Major Aquaculture Taxa
and Their Geographic Importance

INTRODUCTION

There is an estimated total of 10 to 30 million species of organisms now living on earth. Of these, only about 770,000 plant and 1.5 million animal species are known to scientists. Most of the undescribed species are tropical insects and other small invertebrates. About 96% of the known plants and 85% of the animals live in terrestrial environments. Of these, only 3,000 plant species are cultivated and very few have great importance. Very few animals are cultivated. The bulk of our plant food comes from only seven species: wheat, rice, maize (corn), potato, barley, sweet potato, and cassava. Similarly, only seven categories of domesticated terrestrial animals provide most of our meat production: swine, cattle, poultry (chicken, duck, and turkey), sheep, goat, buffalo, and horse.

There are about 20,000 species of fish, 200,000 species of aquatic invertebrates, and 14,000 species of aquatic plants. Very few of these are subjects of aquaculture, but relatively speaking, a greater diversity of aquatic species than of terrestrial species is cultivated.

About 500 aquatic species have been cultivated commercially, although very few account for significant tonnages. A recent survey showed that at least 314 finfishes, 74 crustaceans, 69 mollusks, 43 algae, 13 angiosperms (including rice), 12 sponges, nine amphibians, four reptiles, three rotifers, two annelids, two mammals, and one echinoderm are subjects of aquaculture.

Most of the important cultured species are poikilothermic and carnivorous. Most species are virtually identical to wild stocks, and in many aquacultured species, seed stock is actually obtained from the wild. Development of domesticated stocks has been a recent phenomenon. Some feel that common carp and rainbow trout are the only truly domesticated food fishes in the same sense that terrestrial livestock is domesticated.

There have been several highly visible cases where hardy, easy-to-cultivate species have been transplanted around the world. Tilapias, common carp, rainbow trout, and several salmon species have become such cosmopolitan species. Nonnative oysters such as the Japanese oyster have been introduced into North American and European waters. More than 100,000 metric tons of this oyster are cultivated annually in France while only a few thousand tons of the indigenous European flat oyster are now produced. Freshwater crayfish species have been translocated from the U.S.A. to Europe, Africa, Asia, Central America, the Caribbean, the Pacific, and South America. The purpose of this section is to provide an overview of important indigenous and introduced aquacultural taxa in various parts of the world.

COSMOPOLITAN AQUACULTURE TAXA

Some genera of fish and shellfish dominate the international aquaculture scene. Most of them are cultivated on every continent except Antarctica. They include the rainbow trout, the Atlantic salmon, Pacific salmons (genus *Oncorhynchus*), Chinese carps including the common carp, various tilapias, penaeid shrimps,[1] and oysters (genus *Crassostrea*). One freshwater crayfish species, *Procambarus clarkii*, also has an international distribution, although it has not yet been introduced to the Australian continent.

Tilapia farming started in Egypt almost 4,000 years ago. Some people believe that tilapia culture in Africa is older than the Chinese carp culture in China. The tilapias are a group of freshwater but euryhaline herbivorous and omnivorous fishes that care for their young. Their name is derived from an African bushman word sim-

---

1. The United Nations Food and Agriculture Organization (FAO) now classifies marine penaeid crustaceans as "shrimps" and "freshwater" crustaceans of the genus *Macrobrachium* as "prawns." This protocol is used herein. It should be recognized, however, that the term "prawn" is commonly used throughout the world to describe *all* shrimp-like crustaceans.

ply meaning "fish." There are about 70 species, most of them native to western Africa. Some species can tolerate full strength sea water. There are three important genera including *Tilapia* (nest guarders), *Sarotherodon* (mouth brooders), and *Oreochromis* (mouth brooders) (Table 8.1. and Illustration 8.1).

Oreochromis mossambicus, the Java or Mozambique tilapia, was the first species to become widely known outside of Africa. The most popular species today may be *Oreochromis niloticus*, the Nile tilapia, but other species and their hybrids are gaining recognition due to some special characteristics. One of the best-known hybrid strains is the red tilapia that originated in Taiwan. The tilapias are rapid growers with tasty flesh. Two major disadvantages of these fishes are early, uncontrolled breeding that leads to overpopulation and stunting; and poor tolerance of temperatures below 10 to 15°C. Tilapia production approached 400,000 metric tons in 1990. The most important species are presented in Table 8.1.

Carp culture had its origins in Asia between 3,000 and 4,000 years ago. There are many different species of carps. The Chinese

ILLUSTRATION 8.1. Blue tilapia, *Oreochromis aureus*, a low tropic level, tropical/semi-tropical mouth brooder. Credit: Jay Huner.

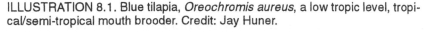

TABLE 8.1. The more important cultured tilapias. Some authorities, including the American Fisheries Society, place all of these species into one genus, *Tilapia*.

Blue tilapia	*Oreochromis aureus*
Wami tilapia	*O. hornorum*
Longfin tilapia	*O. macrochir*
Java (Mozambique) tilapia	*O. mossambicus*
Nile tilapia	*O. niloticus*
Red tilapia	*O. niloticus* x *O. mossambicus*
St. Peter's fish	*Sarotherodon galilaeus*
Redbreast tilapia	*Tilapia rendalli*
Zill's tilapia	*T. zillii*

species are the best-known but the Indian species are now assuming more prominence. Carp culture is very productive because it is based on the idea of combining different species using different feeding niches, that is, polyculture. Usually this is further integrated with terrestrial farm production by the use of animal and plant wastes to enrich ponds. Annual combined production of all Chinese carps exceeds 3 million metric tons in 1990. The important Chinese carp species, their habitats, and their feeding niches are presented in Table 8.2.

The proportion of each species stocked in each pond or lake has been based on trial and error over many centuries. The Chinese carps as well as the Indian carps are very often stocked with one or two of the following species: striped mullet (*Mugil cephalus*–detritivore) (Illustration 8.2), milkfish (*Chanos chanos*–detritivore), bream (*Parabramis pelinensis*–omnivore), Java tilapia (*Oreochromis mossambicus*–detritivore) and wuchang fish (*Megalobrama amblycephala*–herbivore).

The common carp (Illustration 8.3) and rainbow trout are probably the most cosmopolitan finfishes, that is, introduced and cultured in the most countries. However, rainbow trout cannot tolerate warm waters; in the tropics and subtropics they can only be cultured in mountain areas. Production of the two species exceeded 1,113,000 and 270,000 metric tons, respectively, in 1990.

TABLE 8.2. Chinese carp species: habitat and feeding niches.

English name	Latin name	Water level inhabited	Principal source of food
Grass carp	*Ctenopharyngodon idella*	Middle	Macrophytes
Silver carp	*Hypophthalmichthys molitrix*	Upper	Phytoplankton
Bighead carp	*Hypophthalmichthys** *(Aristichthys) nobilis*	Middle	Zooplankton
Mud carp	*Cirrhina molitorella*	Bottom	Omnivore Detritus
Black carp	*Mylopharyngodon piceus*	Bottom	Benthic snails, clams and worms
Common carp	*Cyprinus carpio*	Bottom	Omnivore
Crucian carp	*Carassius carassius*	Middle	Invertebrates

Natural fisheries for Atlantic salmon (Illustration 8.4) have declined to historically low levels of catch. However, aquaculture of this species is a major industry in Scandinavia and northwestern Europe. It is also being cultivated on both coasts of North America and in the Southern Hemisphere, but Europe is the dominant producer. Worldwide production of Atlantic salmon in 1990 exceeded 230,000 metric tons (about 160,000 metric tons in Norway alone) compared to natural fisheries of that species which account for less than 15,000 metric tons annually.

The Pacific salmon species are very prolific, but natural harvest is highly variable from year to year. In Alaska, which accounts for most of the U.S. catch, 20 million salmon were landed in 1967, and 144 million in 1985. In 1989 the world catch was more than 700,000 metric tons. The five species of Pacific salmon listed in Table 8.3 have different characteristics, and only chinook (*Oncorhynchus tshawytscha*) and coho (*Oncorhynchus kisutch*) are considered to meet the same quality standards that are applied to Atlantic salmon. The Pacific species are cultured for ranching both in their

---

* According to FAO Fisheries Circular No. 815 Revision 4, the genus has changed.

ILLUSTRATION 8.2. Striped mullet, *Mugil cephalus*, a very important, low tropic level estuarine species distributed around the world. Young are captured in estuaries. Hatchery production is not practical. Credit: J. W. Avault, Jr., Louisiana Agricultural Experiment Station.

native North Pacific and in the Southern Hemisphere (Chile). The coho salmon has also been cultured in cages in Europe. Aquaculture production of Pacific salmon for food exceeded 60,000 metric tons in 1990. However, it should be noted that many of the Pacific salmon captured in natural fisheries in the North Pacific were actually cultured to the smolt stages by fisheries cooperatives and governmental fish hatcheries. These represent major aquacultural enterprises in themselves even though they are not private or for profit.

The milkfish (Illustration 8.5) is one of the most important euryhaline finfish species in Asia. Over 400,000 metric tons were grown in 1990. There is major production in Taiwan, the Philippines, and Indonesia. It is frequently cultivated with penaeid shrimps and other finfish species in estuarine ponds. The stocking of fingerlings in pens in Laguna de Bay near Manila is well-known. One third of the 90,000-ha (hectare) lake is occupied by net pens

ILLUSTRATION 8.3. Large-scaled variety of common carp, *Cyprinus carpio*, perhaps the most important cultivated freshwater fish. Credit: J. W. Avault, Jr., Louisiana Agricultural Experiment Station.

ranging in size from 1 to 100 ha. However, most milkfish culture is done in open water in shallow, brackish-water ponds in which the fish feed mainly on rich growths of benthic algae.

The farming of penaeid shrimps (Illustration 8.6) is very important due to an apparently insatiable international demand for marine shrimps. Pond production of penaeids grew from negligible amounts in the 1970s to 690,000 metric tons, about 28% of the

ILLUSTRATION 8.4. Atlantic salmon, *Salmo salar*, an anadromous species most commonly cultivated in net pens in temperate waters in the Northern and Southern Hemispheres. It is also the subject of ocen ranching, especially in the northern Atlantic Ocean. Credit: Hans Ackefors.

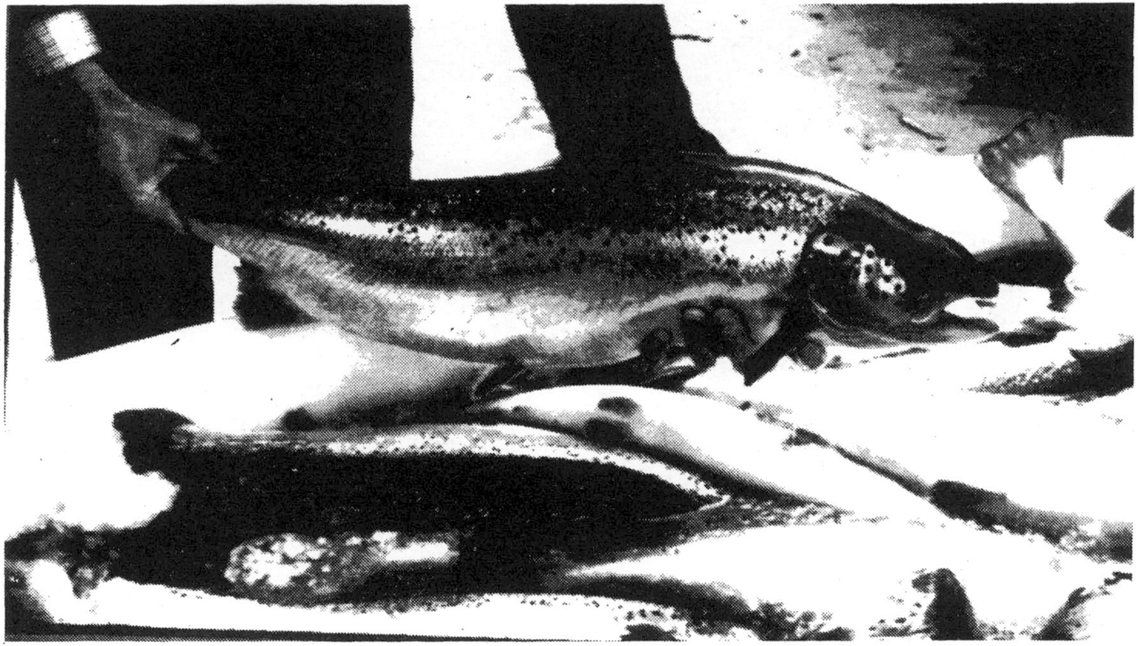

TABLE 8.3. The five most important species of Pacific salmon and significant characteristics.

Common name	Scientific name	Freshwater phase	Maturity age	Max. weight
Chinook salmon (or King salmon)	*Oncorhynchus tshawytscha*	6 months	2-7 years	45 kg
Coho salmon (or Silver salmon)	*O. kisutch*	12 months	2-4 years	15 kg
Sockeye salmon (Red salmon)	*O. nerka*	2-3 years	4-5 years	10 kg
Chum salmon (Dog salmon)	*O. keta*	6-12 months	2-6 years	20 kg
Pink salmon (Humpback salmon)	*O. gorbuscha*	6-12 months	2 years	4-5 kg

world harvest of shrimp, in 1991. This is one of the fastest-growing activities in aquaculture, with a growth rate during the 1980s that was in excess of 32% annually. This rapid growth rate can be attributed to several major breakthroughs in production technology, including nutrition, pond management, and especially hatchery production of larval and post-larval shrimp for stocking. Shrimp farming is practiced in over 40 countries where there is an estimated total of 994,000 hectares (2.45 million acres) in production. Six countries–China, Indonesia, Thailand, Ecuador, Philippines and India–produce over 75% of cultured shrimp. The most important species and their natural distributions are listed in Table 8.4.

The tropical and subtropical shrimps of the genera *Penaeus* and *Metapenaeus* are among the most financially attractive aquacultural species in countries with a suitable climate. In Asia, where over 80% of the world's cultured shrimp are produced, three species–*P. chinensis, P. monodon,* and *P. merguiensis*–account for over 75% of the production, with the remainder distributed among nine to ten other species. A similar situation exists in the Western Hemisphere, with *P. vannamei* accounting for about 90% of the production and

ILLUSTRATION 8.5. Milkfish, *Chanos chanos,* a very important, low tropic level estuarine species in the Orient. Young are captured in estuaries. Hatchery production is not practical. Credit: J. W. Avault, Jr., Louisiana Agricultural Experiment Station.

nine to ten others being grown more or less experimentally. Most species are usually cultivated in their own native areas, but many trials are being made to import nonnative species that may have good culture potential. For example, the Asian *Penaeus japonicus* (which has a wider temperature tolerance than many species) has been introduced into Mediterranean lagoons and is cultured in Brazil; and the Eastern Pacific *Penaeus vannamei* (which shows good growth characteristics even under crowded conditions) is being cultivated in western Atlantic areas.

Penaeid shrimps have complex life cycles that include both off-shore areas and coastal areas such as marshes, lagoons, and mangrove swamps. Under natural conditions, spawning usually takes place in open waters at oceanic salinities. Post-larvae are brought to nursery areas near shore by tides and associated currents. After a few months of growth, they migrate as subadults back to the sea to mature, reproduce, and die. The weight of the different species as

ILLUSTRATION 8.6. Marine penaeid shrimp, *Penaeus* spp. Cultivation of penaeid shrimps constitutes the most important form of marine crustacean culture in the world. Credit: Jay Huner.

TABLE 8.4. Some important cultured penaeid shrimps and their natural distributions.

Common name	Scientific name	Distribution
Chinese white	*Penaeus chinensis*	Yellow and East China Seas
Mexican brown	*P. californiensis*	E. Pacific, California to Peru
Red	*P. brevirostris*	E. Pacific, Mexico to Peru
Whiteleg	*P. vannamei*	E. Pacific, Mexico to Peru
Blue	*P. stylirostris*	E. Pacific, Mexico to Peru
Indian white	*P. indicus*	Indo-West Pacific, from E. Africa to India, China and Australia
Banana	*P. merguiensis*	Indo-West Pacific, Arabian Gulf to Malaysia, China, and Australia
Giant tiger	*P. monodon*	Indo-West Pacific, from S. Africa to India, Japan, and Australia
Kuruma	*P. japonicus*	Indo-West Pacific, from S. Africa to Korea, Japan, and N. Australia
Brown	*P. aztecus*	Gulf of Mexico and W. Atlantic, from Yucatan to Massachusetts
White	*P. setiferus*	Gulf of Mexico and W. Atlantic, from Yucatan to New York
Pink	*P. duorarum*	Gulf of Mexico and W. Atlantic, from Mexico to Maryland
Caramote	*P. kerathurus*	Mediterranean and E. Atlantic from England to Angola
White	*Metapenaeus dobsoni*	India

adults varies from 10 to 100 g. Under hatchery conditions, the culturist must adjust water salinities, temperatures, and foods to match those in the natural habitats of the developing larvae. The post-larvae and juveniles are then stocked out in ponds where they grow to harvestable size.

Some shrimps are able to spend most of their lives in freshwater, for example, the giant freshwater prawn, *Macrobrachium rosenbergii*, which is native to the tropical and subtropical Indo-Pacific region. This prawn has been the object of aquacultural endeavors around the world, including in temperate climates. World production in 1990 was approximately 23,000 metric tons. In its native habitat it is found in virtually all types of fresh and brackish waters, but larval development requires water of 8 to 22 ‰ salinity. Juveniles migrate upstream two to three months after hatching and spend six to eight months in freshwater before they migrate back to the coast for reproduction. The males reach lengths of 25 cm excluding the long claws. Unfortunately, temperatures below 15°C are invariably fatal, which restricts the growing season in temperate climates. Another problem in temperate climates is that giant prawns can be highly territorial, and this often leads to a preponderance of nonuniform sizes at the end of the short growing season.

Cupped oysters of the genus *Crassostrea* are cultured throughout the world (Table 8.5). Production was nearly 900,000 metric tons in 1990. Most oysters are grown in their native regions, but one species, the Pacific cupped oyster, *Crassostrea gigas* (Illustration 8.7) has had great success outside its indigenous range. It is a hardy species with a rapid growth rate, reaching a weight of 30 to 60 g within six to 18 months, and it accounts for over 80% of the world production of oysters by aquaculture.

Other mollusks important in aquaculture include mussels, clams and cockles, and scallops and pectens. Over 1 million metric tons of marine food mussels, *Mytilus* spp., were produced in 1990. The most important species are *Mytilus edulis* and *M. galloprovincialis*. In 1990, the world production of clams, arkshells, and cockles was about 500,000 metric tons while the world production of scallops and pectens was about 340,000 metric tons.

TABLE 8.5. Cultivated cupped oysters, genus *Crassostrea.*

Oyster Common name	Scientific name	Native area	Cultured area
Portuguese	*Crassostrea angulata*	Portugal, Spain	Portugal, Spain
Sydney rock	*C. commercialis*	Australia, New Zealand	Australia, New Zealand
Slipper	*C. eradelis*	Philippines	Philippines
Pacific cupped	*C. gigas*	Asia	Asia, Europe, North America
Mangrove	*C. rhizophorac*	Caribbean	Caribbean
American	*C. virginica*	U.S. Atlantic	U.S. Atlantic

SPECIES CULTIVATED IN ASIA

The Japanese have approached aquaculture in a scientific way during the past several decades and are recognized as world leaders in the field. Japanese scientists have succeeded in culturing over 60 species of marine finfish. This is admirable considering the difficulties of inducing marine species to spawn and of culturing their larval stages. Among those species, 17 are presently cultivated commercially to marketable or fingerling size for restocking the sea. The Japanese also cultivate great quantities of freshwater finfishes (Illustrations 8.8 and 8.9) and marine mollusks and crustaceans. The Japanese cultured pearl industry is world-famous. Japanese aquaculturists are also culturing turtles successfully. Finally, seaweed cultivation is a major Japanese aquaculture industry (Illustration 8.10). An array of species cultured in Japan is presented in Table 8.6. This table demonstrates the diversity of aquatic species that may be cultivated in any temperate climate.

Within Asia, Taiwan is second only to Japan in applying modern science to develop a major aquaculture industry. Freshwater fin-

ILLUSTRATION 8.7. Cultivated Pacific cupped oysters, *Crassostrea gigas*. Credit: Hans Ackefors.

fishes (Illustration 8.11) are relatively more important than marine species in Taiwan. The freshwater species are the same as those cultivated throughout East and Southeast Asian countries such as the People's Republic of China, Vietnam, the Philippines, Thailand, and Malaysia. A list of Taiwanese commercial aquatic species is presented in Table 8.7. Many additional species are being considered as aquacultural candidates.

We would be remiss if we did not emphasize the major role of the People's Republic of China in its development of the Chinese carp polyculture system, now duplicated, with some modifications, around the world. China may not have developed cultural methods for as many species as neighboring Japan and Taiwan, but by weight, its production of edible aquaculture products far exceeds that of any other nation in the world. About half of all cultured finfish and about a third of the annual world total aquacultural crop are produced in China. Furthermore, it is not only in traditional

ILLUSTRATION 8.8. Yellowtail, *Seriola* sp., one of the most common food fishes cultivated in marine cages in Japan. Fingerlings may be produced in hatcheries, but most are still caught at sea. Credit: Osamu Fukuhara.

IILLUSTRATION 8.9. Eels, *Anguilla* sp., in a transport container. Young are captured in river mouths as no one has yet spawned any *Anguilla* species in captivity. Credit: J. W. Avault, Jr., Louisiana Agricultural Experiment Station.

ILLUSTRATION 8.10. Cultivated algae, Nori (*Porphyra* spp.), grown on nets in a Japanese bay. Credit: Hans Ackefors.

aquaculture that the Chinese excel; in the late 1980s they became the top producer of cultured penaeid shrimp as well.

It would be beyond the scope of this text to go into details regarding the diverse species cultured in all of Asia. The cultivation of some selected species in other countries should, however, be mentioned.

There are several important freshwater fish species cultured in Thailand, including the gourami fish, Sepat Siam (*Trichogaster pectoralis*), the Thai silver carp (*Puntius gonionotus*), and the walking catfish (*Clarias batrachus*). In addition, the marine grouper (*Epinephelus tauvina*) and the marine seabass (*Lates calcarifer*) are important species cultured in estuarine net pens.

The culture of molluskan species is of major importance throughout southeastern Asia. Significant quantities of oysters are cultivated for food. Culture of the scallop, blue mussel, hard clam, and abalone is especially noteworthy. The cultivation of bivalve mollusks for pearl production is a very significant source of foreign

TABLE 8.6. An array of marine (M), freshwater (F), and anadromous (A) species cultivated in Japan.

Common name	Scientific name	Habitat
Finfish		
Yellowtail	*Seriola quinqueradiata*	M
Red seabream	*Pagrus major*	M
Black seabream	*Acantopagrus schegeli*	M
Trout salmon	*Oncorhynchus masu*	A
Amago	*Oncorhynchus rhodurus*	A
Horse mackerel	*Trachurus japonicus*	M
Striped jack	*Caranx delicatissimus*	M
Tiger puffer	*Sphaeroides rubripes*	M
Japanese flounder	*Paralichthys olivaceus*	M
Japanese eel	*Anguilla japonica*	A
European eel	*Anguilla anguilla*	A
Common carp	*Cyprinus carpio*	F
Rainbow trout	*Oncorhynchus mykiss*	F
Ayu sweetfish	*Plecoglossus altivelis*	F
Mollusks		
Scallop	*Chlamys nobilis*	M
Japanese scallop	*Patinopecten yessoensis*	M
Blood cockle	*Anadora granosa*	M
Pacific cupped oyster	*Crassostrea gigas*	M
Portuguese oyster	*C. angulata*	M
Pearl oyster	*Pinctada fucata*	M
Pearl oyster	*Pteria penguin*	M
Abalone	*Haliotis* spp.	M

Common name	Scientific name	Habitat
Crustaceans		
Kuruma shrimp	*Penaeus japonicus*	M
Giant tiger shrimp	*P. monodon*	M
Green tiger shrimp	*P. semisulcatus*	M
Greasyback shrimp	*Metapenaeus ensis*	M
Spiny lobster	*Palinurus japonicus*	M
Reptiles		
Soft-shelled turtle	*Amida japonica*	F
Seaweeds		
"Nori"	*Porphyra* spp.	M
"Aonori"	*Monostroma*	M
	Enteromorpha	M
Wakame	*Undaria pinnatifida*	M
Kombu	*Laminaria japonica*	M

exchange in Asia. In freshwater, species such as *Huyriopsis cumingii* and *Christearia plicata* are grown in China for culturing pearls. Many of these freshwater pearls are produced for medicinal purposes, but they are also becoming popular as jewelry. In Japan, pearl oyster operations are concentrated in coastal waters in the middle and southern parts of the country. The dominant species are *Pinctada fucata* and *Pteria penguin.*

Reptiles including turtles, crocodiles, and alligators are cultivated for meat, skin, and other commercial products in Asia. Frogs, *Rana* spp., are also important species grown for meat.

China, Japan, and Korea are the world's main producers of marine algae. The principal genera are *Laminaria, Gracilaria, Unda-*

ILLUSTRATION 8.11. Grass carp, *Ctenopharyngodon idella*, is an important species in integrated polyculture and aquatic weed control. Credit: J. W. Avault, Jr., Louisiana Agricultural Experiment Station.

ria ("Wakame"), and *Porphyra* ("Nori"). Seed production can be accomplished in land-based operations. Grow-out then takes place in protected coastal waters. In some tidal areas, however, wild seed is still collected and the entire culture cycle takes place in situ.

TAXA CULTIVATED IN EUROPE

Fewer than 30 aquatic species are cultivated in Europe–a very small number compared to that in Asia (Table 8.8); and only a few of these 30 species have any great importance. The dominant finfish species are common carp, rainbow trout, and Atlantic salmon. Molluskan aquaculture is largely limited to Pacific cupped oyster and blue mussel. The production of crustaceans is of minor importance. There is practically no commercial farming of seaweeds.

Common carp is cultivated most intensively in Eastern Europe. Atlantic salmon production takes place mainly in the northwestern coastal countries, especially Norway and Scotland. Rainbow trout are cultivated, to some degree, in all countries. In the Mediterranean countries, mullets, European eel, seabream, and seabass are cultivated on a relatively minor scale. European eel and turbot are also cultivated on a minor scale in high-technology operations with heated "waste" water, or in ambient waters in southwestern Europe. Ocean and lake ranching of salmonids (Atlantic salmon, brown trout, chars, and grayling) and coregonids (whitefishes) has become increasingly important, with fingerlings and smolts being stocked for harvest as adults by commercial and sport fishermen. Pike (*Esox lucius*) and pike-perch (*Stizostedion lucioperca*) fry and fingerlings are cultured in several countries for stocking into natural waters to maintain fisheries there. Aquaculturists are also developing culture methods for cod and halibut as food fishes.

Pacific oyster culture takes place mainly in France while blue mussel (including Mediterranean mussel) production (Illustration 8.12) is important over a wider area including Spain, the Netherlands, France, and Germany. Scallop and clam cultures have recently been developed.

Lobsters (*Homarus vulgaris*) are being reared on a trial basis for restocking French, British, and Norwegian waters. Pilot-scale (intermediate between small-scale experimental and full-scale commercial) culture of several penaeid shrimps is underway in France and Italy. Extensive production of freshwater crayfish is widespread in Europe, with emphasis on the native noble crayfish, *Astacus astacus*, and narrow-clawed crayfish, *A. leptodactylus*, as well as the signal crayfish, *Pacifastacus leniusculus*, and the red swamp crayfish, *Procambarus clarkii*, the latter two introduced from North America.

TAXA CULTIVATED IN NORTH AMERICA

Climates in North America range from tropical to cold-temperate and a wide array of species is cultured (Table 8.9). Among the various regions, aquaculture is most developed in the southern U.S.A. Baitfishes, ornamental fishes, channel catfish, rainbow trout, Atlantic salmon, and Pacific salmon are important cultured fishes. A number of sport species (Illustration 8.13) are cultivated as fingerlings for stocking waters for recreational fisheries. These

TABLE 8.7. The species of freshwater (F), brackish (B), and marine (M) organisms cultivated in Taiwan.

Common name	Scientific name	Habitat
Finfish		
Japanese eel	Anguilla japonica	F
Common carp	Cyprinus carpio	F
Grass carp	Ctenopharyngodon idella	F
Silver carp	Hypophthalmichthys molitrix	F
Big head carp	Hypophthalmichthys nobilis	F
Black carp	Mylopharyngodon piceus	F
Mud carp	Cirrhinus molitorella	F
Crucian carp	Carassius carassius	F
Pond loach	Misgurnus anguillicaudatus	F
Redbelly tilapia	Tilapia zillii	F
Mozambique tilapia	Oreochromis mossambicus	F
Nile tilapia	O. niloticus	F
Blue tilapia	O. aureus	F
Hybrid tilapia	O. mossambicus x niloticus	F
Hybrid tilapia	O. niloticus x aureus	F
Red tilapia	Oreochromis hybrids	F, B, M
Walking catfish	Clarias fuscus	F
Japanese catfish	Parasilurus asotus	F
Thailand catfish	Pangasius sp.	F
Snakehead	Channa maculata	F
Largemouth bass	Micropterus salmoides	F
White fish	Culter erythropterus	F
Rainbow trout	Oncorhynchus mykiss	F
Ayu sweetfish	Plecoglossus altivelis	F
Milkfish	Chanos chanos	F, B, M
Rice-field eel	Fluta alba	F
Wuchang fish	Megalobrama amblycephala	F
Striped mullet	Mugil cephalus	F, B, M
Japanese sea perch	Lateolabrax japonica	M
Barramundi perch	Lates calcarifer	M
Red porgy	Chrysophrys major	M
Mud skipper	Boleophthalmus chinensis	M

Common name	Scientific name	Habitat
Mollusks		
Pacific oyster	*Crassostrea gigas*	M
Hard clam	*Meretrix lusoria*	M
Small abalone	*Haliotrix diversicolor*	M
Freshwater clam	*Corbicula fluminea*	F
Formosa clam	*C. formosana*	F
Blood cockle	*Anadara granosa*	M
Purple clam	*Soletellina diphos*	M
Apple snail	*Ampullarium insularum*	F
Crustaceans		
Giant tiger shrimp	*Penaeus monodon*	B, M
Kuruma shrimp	*P. japonicus*	M
Red tail shrimp	*P. penicillatus*	B, M
Sand shrimp	*Metapenaeus ensis*	B, M
Giant freshwater prawn	*Macrobrachium rosenbergii*	F
Mangrove crab	*Scylla serrata*	B, M
Reptiles		
Soft-shell turtle	*Trionyx sinensis*	F
Siamese crocodile	*Crocodilus siamensis*	F
Estuarine crocodile	*C. porosus*	F, B
Amphibians		
American bullfrog	*Rana catesbeiana*	F
Seaweeds		
Gracilaria	*Gracilaria* spp.	B, M
Nori	*Porphyra* spp.	M

TABLE 8.8. The more important freshwater (F), marine (M), catadromous (C), and anadromous (A) species cultured in Europe.

Common name	Scientific name	Habitat
Finfish		
Common carp	*Cyprinus carpio*	F
Brown trout	*Salmo trutta*	F
Rainbow trout	*Oncorhynchus mykiss*	F, A, M
Atlantic salmon	*Salmo salar*	A
Eel	*Anguilla anguilla*	C
Turbot	*Scopthalmus maximus*	M
Sea bream	*Sparus auratus*	M
Sea bass	*Dicentrarchus labrax*	M
Mullets	*Mugil* spp.	C
Whitefish	*Coregonus* spp.	F
Mollusks		
Pacific cupped oyster	*Crassostrea gigas*	M
Blue mussel	*Mytilus edulis*	M
Mediterranean mussel	*Mytilus galloprovincialis*	M
Carpet shell	*Tapes* spp.	M
Crustaceans		
Noble crayfish	*Astacus astacus*	F
Narrowed-clawed crayfish	*Astacus leptodactylus*	F
Signal crayfish	*Pacifastacus pacifastacus*	F
Red Swamp crayfish	*Procambarus clarkii*	F

species include the walleye and sauger (*Stizostedion* spp.), black basses (*Micropterus* spp.), sunfishes (*Lepomis* spp.), pikes (*Esox* spp.), catfishes (*Ictalurus* spp.), trouts (*Salmo trutta* and *Oncorhynchus mykiss*), chars (*Salvelinus* spp.), striped bass (*Morone saxatilis*) and hybrids, and salmons (*Salmo salar* and *Oncorhynchus* spp.).

The widespread interest in sport fishing in the U.S.A. is reflected by the production of major quantities of baitfishes including golden

ILLUSTRATION 8.12. Cultivated blue mussels, *Mytilus edulis*. Credit: L. O. Loo.

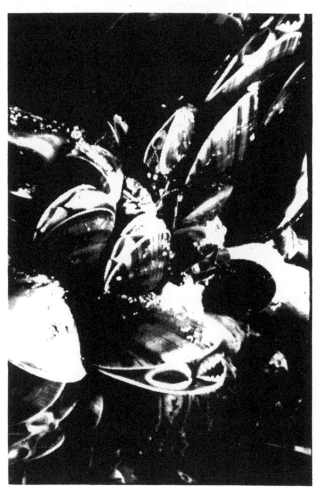

shiner, fathead minnow, white sucker, and goldfish. Channel catfish (Illustration 8.14) culture in the past two decades is a major success story, preceded historically by the rainbow trout (Illustration 8.15) industry which originated at the end of the 1800s. Although their tonnage is low, the value of exotic tropical fishes raised in southern Florida is substantial. The numbers of ornamental species cultivated is far too great to list here. Important shellfish species in the U.S.A.

TABLE 8.9. Important freshwater (F), brackish (B), marine (M), and anadromous (A) species cultivated in North America for food (f), bait (b), stocking (s), or ornamental (o).

Common Name	Scientific Name	Habitat	Utilization
Chinook salmon	*Oncorhynchus tshawytscha*	A	f, s
Coho salmon	*O. kisutch*	A	f, s
Sockeye salmon	*O. nerka*	A	s
Chum salmon	*O. keta*	A	s
Pink salmon	*O. gorbuscha*	A	s
Rainbow trout	*O. mykiss*	A, F	f, s
Atlantic salmon	*Salmo salar*	A, F	f, s
Brown trout	*S. trutta*	F	s
Brook trout	*Salvelinus fontinalis*	F	s
Lake trout	*S. namaycush*	F	s
Striped basses	*Morone* spp.	F, A	f, s
Black basses	*Micropterus* spp.	F	s
Sunfishes	*Lepomis* spp.	F	s
Goldfish	*Carassius auratus*	F	b, o
Golden shiner	*Notemigonus crysoleucas*	F	b
Channel catfish	*Ictalurus punctatus*	F	f, s
Catfishes	*Ictalurus* spp.	F	f, s
Fathead minnow	*Pimphales promelas*	F	b
White sucker	*Catastomus commersoni*	F	b
Koi carp	*Cyprinus carpio*	F	o
Big head carp	*Hypophthalmichthys nobilis*	F	f
Grass carp	*Ctenopharyngodon idella*	F	f, s
Buffalofishes	*Ictiobus* spp.	F	f
Tilapias	*Oreochromis* spp.	F, B	f
Walleye/Sauger	*Stizostedion* spp.	F	f, s
Pikes	*Esox* spp.	F	s

Common Name	Scientific Name	Habitat	Utilization
Tropical fish	Many species	F, B, M	o
Red Swamp crayfish	*Procambarus clarkii*	F	f, b
White River crayfish	*P. zonangulus*	F	f, b
Papershell crayfish	*Orconectes immunis*	F	b
Whiteleg shrimp	*Penaeus vannamei*	M	f
Blue shrimp	*P. stylirostris*	M	f
Blue mussel	*Mytilus edulis*	B, M	f
American oyster	*Crassostrea virginica*	B, M	f, s
Pacific cupped oyster	*C. gigas*	B, M	f
Hard clams	*Mercenaria* spp.	M	f, s
Abalone	*Haliotis* spp.	M	f, s
American alligator	*Alligator mississippiensis*	F	f, skins
Red-eared slider	*Trachemys scripta*	F	o

ILLUSTRATION 8.13. Largemouth bass, *Micropterus salmoides*, is a most important freshwater sportfish in North America. Credit: Jay Huner.

ILLUSTRATION 8.14. Channel catfish, *Ictalurus punctatus*, is the dominant cultured foodfish in the U.S.A. Credit: Jay Huner.

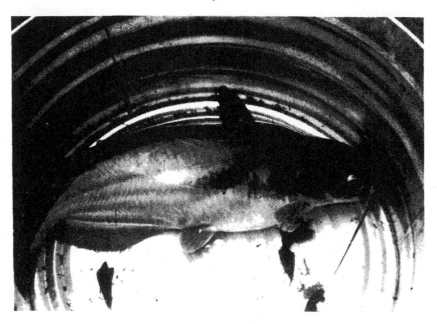

include the red swamp crayfish *Procambarus clarkii* (Illustration 8.16), cultivated extensively in the southern states, oysters (*Crassostrea* spp.) cultivated on all coasts, and the blue mussel, primarily in the Northeast. More effort is being devoted to developing aquaculture of manila clams (*Venerupis japonica*) in the Pacific Northwest and hard clams (*Mercenaria* spp.) along the Atlantic coast. Pioneer work with abalone (*Haliotis* spp.) was done in California. Culture of hybrid striped bass (*Morone chrysops* x *saxatilis*) expanded in recent years from experimental systems to speculative commercial ones. If good markets can be developed, hybrid striped bass culture may become an economic success. This is in contrast to a slow progress for shrimp culture in the U.S.A. Substantial funds have been invested in developing both penaeid and macrobrachium shrimp culture projects in the southern states where, thus far, real successes have eluded investors.

ILLUSTRATION 8.15. Rainbow trout, *Oncorhynchus mykiss* (formerly *Salmo gairdneri,*) is an anadromous temperate species cultivated for food and sportfishing around the world including the higher elevations in the tropics. Large fish is a mature 1 kg foodfish. Smaller fish is a 100 gram smolt suitable for moving from freshwater to saltwater. Credit: Jay Huner.

Alligator (*Alligator mississippiensis*) (Illustration 8.17) is being cultured more than ever in the southern states. Both meat and skins are sold for high prices. Under current practices, the eggs are collected from the wild and hatched under controlled conditions. The young alligators are kept in shallow pools located in heated buildings where they can grow all year long and attain much greater size than wild ones. A proportion of the cultured yearlings must be returned to wildlife officials who restock them in the wild.

The red-eared slider (*Chrysemys scripta*), a pond turtle, is cultivated for the aquarium market. The U.S. market for these animals vanished in the early 1970s after adverse publicity concerning them carrying salmonella bacteria led to a ban on domestic sales. Although new technology has been developed to produce salmonella-free pet turtles, a domestic market has not been reestablished. However, newly hatched turtles are popular pets in several countries and most of the turtles hatched in the U.S.A. are exported.

In Canada, aquaculture production is based on salmonids, blue mussels, scallops, and oysters. Salmon culture (Pacific and Atlantic) is developing very rapidly on the Pacific coast, where there is also some culture of Pacific oyster and blue mussel. Most commercial freshwater aquaculture is limited to rainbow trout. Atlantic salmon and blue mussel are also being cultivated on the Atlantic coast. It should be noted that Mexico has embarked upon an aggressive aquaculture program involving both cash species such as penaeid shrimps and food species such as tilapias and various carps.

TAXA CULTIVATED ELSEWHERE

Aquaculture is not especially well-developed in Africa, Australia, New Zealand, and Latin America, even though there are major

ILLUSTRATION 8.16. A large red swamp crayfish, *Procambarus clarkii*, about 50 grams, facing a wild-caught American lobster, *Homarus americanus*. Methods to cultivate clawed lobsters such as *H. americanus* are well-developed but their culture remains unprofitable. Credit: Jay Huner.

ILLUSTRATION 8.17. American alligators, *Alligator mississippiensis*, in a special environmental chamber to maintain high temperatures. Alligators and crocodiles are being cultivated around the world for their valuable hides and secondarily for their meat. Credit: Jay Huner.

opportunities in those lands. Important aquaculture industries have developed in South America, including penaeid shrimps, especially the whiteleg shrimp; *Penaeus vannamei*, mainly in Ecuador, but to a lesser extent in Colombia and Panama; and Atlantic salmon in Chile. Sea pen culture of Atlantic salmon and rainbow trout is becoming important in the Australian island state of Tasmania. Elsewhere in Australia, the northwest coast is the site of a well-established cultivated pearl industry, and the southeast coast and Tasmania have important oyster culture industries. New Zealand has recently developed a significant greenlip mussel industry. Finfishes, primarily tilapias, are being cultivated throughout sub-saharan Africa but mainly at the subsistence or cottage industry levels.

In the Middle East, Israel practices intensive polyculture using tilapias and Chinese carps. It produces enough of these fishes to supply all its own domestic fish needs.

OTHER SPECIES

The list of aquatic species with cultural potential is long. The FAO recently listed several hundred species reported to be cultivated throughout the world. Many more African, Australian, North American, South American and Central American species have yet to be examined for aquaculture potential. Thus, the future for aquaculture development is great.

A major problem in aquacultural development is that exotic species introduced into an area for aquacultural purposes can escape and establish themselves in the natural systems. This invariably leads to ecological changes that are most controversial. Thus it is wise, if at all possible, to determine whether native species have aquaculture potential before translocating nonnative species. In cases where nonnative species are used for aquaculture, rigorous precautions should be taken to prevent their escape and establishment of breeding populations in the wild. For example, in the U.S.A., sterile, triploid (three sets of chromosomes), nonnative grass carp are approved for use in aquatic weed control in most states, whereas the fertile, diploid (normal two sets of chromosomes) grass carp are illegal. Should a nonnative species become established in the wild, it is most important that local regulatory agencies make every effort to assure that feral stocks are exploited to the benefit of the peoples of the region. This is mainly an educational endeavor in which information is circulated on how to catch, dress, store, market, and cook the new species.

Chapter 9

Global Production of Aquaculture Products

INTRODUCTION

In 1966, global aquaculture production was estimated at 1 million metric tons. By 1989, in just 23 years, total world aquaculture production was more than 14 times that figure. During the same period, yield from capture fisheries increased only about 53%. Most estimates of future capture fisheries predict a plateau of about 100 million metric tons well into the next century. This is only 13% above the current yield. Predictions for aquaculture production, however, forecast continued growth, with most estimates ranging from 20 to 25 million metric tons by the year 2000, and up to 60 million metric tons by 2025. In the keynote address to the 1991 World Aquaculture Conference, Michael B. New reviewed the history of aquaculture predictions and noted that in the past many have been overly conservative.[1]

The fact that the yield from capture fisheries is reaching a plateau is surprising to many people. Technological advances in capture fisheries during the past 50 years have included radical improvements in ships, engines, navigation, fishing gear, artificial netting materials, refrigeration and preservation, fish detection devices, electronics, communications, and use of aircraft to locate fish. Growth in capture fisheries since 1950 can be divided into three

---

1. For example, in 1985 several authorities predicted that world shrimp culture production would reach 250 million metric tons by 1990. In reality, production levels more than twice this number were reached by 1989. Michael B. New. 1991. "Turn of the millennium aquaculture." *World Aquaculture* 22(3): 28-49.

stages. First, between 1950 and the end of the 1960s, landings of fish and shellfish increased steadily at a rate of about 5% annually, or from about 22 million metric tons to about 65 million metric tons. In the second phase, during the 1970s, the annual increase leveled off to only about 1%, while in the third phase, during the 1980s, landings again increased more rapidly, by an average of 3% annually, reaching 88.5 million metric tons (not including marine mammals or seaweeds) in 1989. It is now thought that growth will again level off. By the 1990s, the new technologies have been applied to most of the world's sustainable fishery resources (and some that are not sustainable, as well). Therefore, there is little potential for future growth of capture fisheries beyond the predicted 100 million metric tons annually.[2]

On the other hand, the annual average increase in aquaculture production from 1984 through 1989 was 6.8%, and growth is expected to continue through the year 2000 at an annual rate of about 5% (Table 9.1). Aquacultural production in 1984 was 10.1 million metric tons, and the projected figure for the year 2000 is 25 million metric tons (includes about 4 million tons of seaweeds). This is somewhat below, but reasonably close to, the production target of 30 million metric tons that FAO set in the mid-1970s as a goal for the turn of the century. Still, world aquaculture production continues to grow in many areas. In recent years there has been a conspicuous increase in the culture of salmon in Europe and the Americas, of tilapia in Asia, of shrimps in Asia and South America, and of catfishes and crayfishes in North America. The production of mollusks and seaweeds has been relatively static.

The total world supply of fish, shellfish, and seaweeds in 1989 was 103.8 million metric tons, of which some 86.5% came from capture fisheries and 13.5%, or 14.0 million metric tons, from aquaculture. Based on the 1989 figures in Table 9.2, the yield from cultural activities consisted of 52.3% finfish, 21.6% mollusks, 4.4% crustaceans, 21.4% seaweeds and algae, and 0.3% other species.

Comparing capture fisheries and cultural activities, 8.4% of the finfish was cultivated, 56.8% of the mollusks, excluding squids and

---

2. Source: FAO. 1989. *Review of the State of World Fishery Resources*. FAO Fisheries Circular No. 710 rev. 6.

octopuses (cephalopods), 13.6% of the crustaceans, and 69.1% of the seaweeds and algae. If seaweeds and algae are removed from the calculation, aquaculture production was 11.0 million metric tons, which was 11.1% of the world animal-based fisheries yield of 99.5 million metric tons (Table 9.2).

According to the latest estimate from FAO, published in 1991, not less than 83% of the aquaculture production in 1989 came from Asia (Table 9.3). Thus, both well-developed industrial societies and underdeveloped agrarian societies outside of Asia have yet to realize their full aquacultural potentials.

By dividing the present aquaculture production into various commodity groups, it is possible to obtain a better global picture of aquaculture development (Table 9.4). It is obvious that finfishes dominate production in most regions except in South America and Europe. Finfish account for 52.3% of total world aquaculture production. Mollusks and algae account for 21.6% and 21.4%, respectively. Bivalve mollusk production, including oysters, is greatest in

TABLE 9.1. The total yield of various aquaculture groups during 1984-89 with projections for the year 2000 (millions of metric tons).

Aquaculture Group	Annual World Production			
	1984[a]	1987[b]	1989[b]	2000[c]
Finfish	4.335	6.550	7.327	9.67
Mollusks	2.045	2.719	3.033	7.06
Crustaceans	0.228	0.589	0.612	1.27
Seaweeds	3.446	3.023	2.998	4.20
Others	0.021	0.028	0.048	
Total	10.075	12.909	14.018	22.20[d]

[a] Source: M. B. New. 1991. *World Aquaculture* 22(3).
[b] Source: FAO. 1991. Fisheries Circular No. 815 rev. 3.
[c] Source: FAO. 1987. ADCP/REP/87/25.
[d] Many recent authors now give a figure of 25 million metric tons.

TABLE 9.2. The total world supply of fish, shellfish, and seaweeds/algae in 1989 from capture fisheries and aquaculture. Percentages refer to total yield for each commodity group.

Commodity Group	Capture fisheries[a]		Aquaculture[b]		
	million metric tons	%	million metric tons	%	Total yield
Finfish	79.488	91.6	7.327	8.4	86.815
Crustaceans	3.891	86.4	0.612	13.6	4.502
Mollusks (excl. Cephalopods)	2.309	43.2	3.033	56.8	5.342
Cephalopods	2.537	100.0	0.0	0.0	2.537
Others	0.290	85.7	0.048	14.3	0.338
Seaweeds	1.342	30.9	2.997	69.1	4.340
Total	89.856	86.5	14.018	13.5	103.874

[a]Source: FAO. 1991. Fishery statistics: catches and landings. FAO Yearbook 1989.
[b]Source: FAO. 1991. Aquaculture production (1986-1989). FAO Fisheries Circular no. 815 revision 3.

Asia and Europe. Almost all world seaweed cultivation (98.7%) is in Asia.

Production of crustaceans is comparatively small (4.4%); however, the commercial value of cultured crustaceans is great, estimated at 20% of the total sales of all cultured aquatic species. Most crustaceans are cultivated in Asia, but South and North America are becoming important centers as well.

About 46% of world aquaculture production comes from freshwater areas, 7.5% from brackish waters, and 46.5% from marine waters (Table 9.5). The bulk of marine production is made up of mollusks (clams, cockles, oysters, mussels, and scallops) and seaweeds. Most of the freshwater production consists of finfishes. Most crustacean production takes place in brackish waters.

In 1989, the total yield of all fisheries products in freshwater was 13.8 million metric tons and in marine areas 85.8 million metric

TABLE 9.3. The total yield of aquaculture in various regions of the world in 1989 (Source: FAO. 1991. Fisheries Circular 815 rev. 3).

Region	1000 metric tons	Percentage
Africa	95	0.7
North America	462	3.3
Central America[a]	90	0.6
South America	151	1.1
Asia[b]	11,647	83.1
Europe	1,177	8.4
Oceania	42	0.3
USSR (former)	354	2.5
Total	14,018	100.00

[a]includes Caribbean and Mexico
[b]includes Near East

tons excluding seaweeds.[3] This means that 47% of the freshwater harvest came from aquacultural endeavors while the corresponding figures for marine and brackish waters was 5%. If seaweeds are included, the marine and brackish aquaculture harvest increases to 8% of the total marine fisheries yield.

According to FAO, the People's Republic of China was the world's major producer of cultured aquatic products in 1989. Nearly 6.6 million metric tons were produced, about 47% of the world's aquaculture production. World aquaculture production in 1989 is presented in Table 9.6.

In terms of farming practices, the leading form of aquaculture production during the mid-1980s was, by far, the use of ponds and tanks (41%). Mollusk culture on bottom (18%) and off bottom (7%) accounted for most of the remaining identified production (Table 9.7). Although no farming practice data have been published in

---

3. Source: FAO. 1991. *Fishery Statistics: Catches and Landings*, FAO Yearbook 1989.

TABLE 9.4. Production of aquaculture products in 1989 by commodity groups and regions (Source: FAO. 1991. Fisheries Circular 815 rev. 3).

Metric Tons

	Finfish	Crustaceans	Mollusks	Seaweeds	Others	Total
Africa	91,346	138	2,971	0	88	94,543
North America	307,577	30,841	123,287	0	0	461,705
Central America[a]	28,745	8,095	53,466	30	4	90,340
South America	30,202	81,108	3,921	36,150	0	151,381
Asia[b]	5,997,564	487,562	2,175,408	2,957,802	48,218	11,646,625
Europe	535,965	3,067	637,743	0	2	1,176,777
Oceania	4,969	1,058	35,833	445	0	42,305
USSR (former)	350,785	0	310	3,070	0	354,165
Total	7,327,153	611,869	3,033,009	2,997,471	48,339	14,017,841
Percentage	52.3	4.4	21.6	21.4	0.3	100.0

[a]includes Caribbean and Mexico
[b]includes Near East

recent years, production practices have remained similar, with the exception of an increase in cage production of marine salmonids to about 2% of totals.

The important aquatic taxa and their production figures for 1989, excluding seaweeds, are shown in Table 9.8. The overwhelming predominance of the Chinese carps is clear.

REGIONAL AQUACULTURE PRODUCTION

Asia is the primary producer of cultured aquatic species. China, Japan, the Republic of Korea (South Korea), and the Philippines account for nearly 81% of the total Asian harvest in 1989. With the addition of Indonesia, India, Democratic People's Republic of Korea (North Korea), Thailand, Bangladesh, and Vietnam, ten countries account for 97% of the Asian production. This corresponds to over 11 million metric tons.

TABLE 9.5. Aquaculture production in 1989 by environment and region (1,000 metric tons) (Source: FAO. 1991. Fisheries Circular 815 rev. 3).

| | Metric Tons | | | | |
	Marine	Freshwater	Brackish	World	%
Finfish	522,207	6,374,011	400,935	7,327,153	52.3
Crustacea	38,125	57,141	516,603	611,869	4.4
Mollusca	2,896,114	9,955	126,940	3,033,009	21.6
Seaweeds	2,988,813	11	8,647	2,997,471	21.4
Others	46,798	1,224	317	48,339	0.3
Total	6,522,057	6,442,342	1,053,442	14,017,841	
Percentage	46.5	46.0	7.5		

Asia also leads the world in aquaculture production on an annual per capita basis. Excluding the harvest of seaweed, the 1989 production of aquatic animals by culture was about 2.8 kg per capita in Asia compared to 2.1 kg per capita for the world. Since the per capita yield of wild stocks in Asia, 11.8 kg per capita, is substantially less than the world average, 17.0 kg per capita, in Asia aquaculture provides a much higher portion (19%) of total fishery products (without seaweeds) than the world average (11%). If seaweeds are included in this analysis, the predominance of Asian aquaculture becomes even more pronounced, since nearly 70% of seaweeds are produced by culture, and almost 99% of cultured seaweeds are produced in Asia.

China has the largest production of freshwater animal species, with about 4 million metric tons per year. China also leads in marine animal species with about 1.2 million metric tons. Japan produces about 750,000 metric tons of marine animal species. The Philippines and Indonesia culture more than 300,000 metric tons of milkfish. Three countries–China, Japan, and the Republic of Korea–produce nearly 1.8 million metric tons, about 82%, of the molluskan species cultured annually in Asia. In total, Asia produces nearly 72% of the world's cultured shellfishes.

Culture of penaeid shrimps is expanding in Asian countries. As recently as 1981, aquaculture of marine shrimps was in its infancy, and only about 2% of shrimps were cultured. But by 1991, 28% of

TABLE 9.6. Aquaculture production of the 16 leading countries. These 16 countries produced over 91% of the world supply of aquaculture products in 1989. (Source: FAO. 1991. Fisheries Circular 815 rev. 3).

Country	Metric Tons	Percent of World Total
China	6,557,785	46.8
Japan	1,365,254	9.7
Republic of Korea	859,773	6.1
Philippines	629,325	4.5
Indonesia	497,120	3.6
India	490,000	3.5
U.S.A.	443,359	3.2
USSR (former)	354,165	2.5
Dem. People's Rep. of Korea	341,470	2.4
France	228,894	1.6
Spain	219,503	1.6
Thailand	215,764	1.5
Bangladesh	162,630	1.2
Vietnam	146,700	1.1
Italy	136,925	1.0
Norway	118,810	0.9

the world supply was cultured. Over 80% of farmed shrimp are produced in Asia, with three countries–China, Indonesia, and Thailand–yielding over 57% of the world's cultured shrimp.[4] The rapid development of shrimp culture has greatly impacted world shrimp marketing and pricing structure, and this, in turn, has led to economic difficulties for traditional artisanal shrimpers.

Asia is the major producer of seaweeds in the world, and seaweeds constitute about 25% of Asian aquaculture production.

---

4. Source: Rosenberry. 1991. "World shrimp farming 1991." *Aquaculture Digest.*

TABLE 9.7. Aquaculture production by farming practices in 1985 (% of total excluding seaweeds). (Source: Nash. 1988. Aquaculture communiques. *Journal World Aquaculture Soc.* Vol. 19(2): 51-58.)

Practice	Percentage
Ponds and Tanks	41
Enclosures and Pens	3
Cages	<1
Raceways and Silos	1
Barrages	<1
Other Methods	<1
Mollusks on Bottom	18
Mollusks off Bottom	7
Unspecified	29

China, Japan, the Republic of Korea, and the Philippines are the main producers.

Japan is the leading nation in the world in marine finfish production with about 0.2 million metric tons annually. Japan also emphasizes the enhancement of fish and shellfish species for fisheries by stocking numbers of seed animals into the sea. There is significant hatchery production of seed stock for Kuruma prawn, Yoshi shrimp, Gazami crab, abalone, Japanese scallop, and sea urchins.

The Japanese have a very diversified aquaculture industry. However, as in most countries, a few species are more important than others. The production of yellowtail (also known as buri or Japanese amberjack) has grown rapidly, as has the production of red seabream (madai). The latter species was the object of intense research for many years before the mass rearing of juveniles was accomplished. Commercial production dates only from 1970. Japanese production of oysters is also very impressive at just over 250,000 metric tons annually. Only the macroalgae nori is produced in greater quantities at about 400,000 metric tons annually. Production data for the most important species cultured for food are presented in Table 9.9.

TABLE 9.8. Production of important aquaculture species in 1989 (metric tons) (Source: FAO. 1991. Fisheries Circular 815 rev. 3).

Species	Production	Areas with ≥ 2,000 metric tons annually
Common carp	987,682	China, USSR, Indonesia, Poland, Germany, Czechoslovakia, Japan, Hong Kong, former Yugoslavia, Bulgaria, Israel, Korea, Iran, Iraq, Nigeria, France, Thailand, Turkey
Silver carp	1,432,718	China, former USSR, Hungary
Bighead carp	641,425	China, Hungary
Grass carp	946,963	China
Milkfish	334,478	Philippines, Indonesia
Tilapias	325,562	China, Indonesia, Philippines, Thailand, Cuba, Malaysia, Jamaica, Japan, Nigeria, Israel, Mexico
Rainbow trout	244,898	Italy, Denmark, France, USA, Germany, Japan, Finland, Spain, UK, Sweden, Austria, Canada, Chile, Norway
Channel catfish	183,729	USA
Atlantic salmon	168,290	Norway, UK, Faeroes, Canada, Ireland, USA
Pacific salmon	108,398	USA, Japan, Canada, Chile
River eels	90,101	Japan, Italy
Japanese amberjack (Yellowtail)	154,733	Japan
Freshwater prawn	20,929	Thailand
Red Swamp crawfish	32,690	USA, Spain, People's Republic of China
Fleshy prawn	175,234	China
Giant tiger prawn	142,653	Thailand, Philippines, Indonesia
Whiteleg shrimp	79,157	Ecuador, Peru, Panama
Banana shrimp (prawn)	31,583	Indonesia, Philippines, Thailand
Other shrimps	81,010	Vietnam, Bangladesh, Indonesia, Thailand, Japan, Colombia
Pacific cupped oyster	745,836	Japan, Korea, France, China, USA, Canada

Species	Production	Areas with ≥ 2,000 metric tons annually
Japanese scallop	309,820	Japan, China
Ark clams	16,947	Korea
Blood cockle	93,092	Malaysia, China, Korea, Thailand
Japanese (Manila) clam	154,940	China, Korea
Razor clams	138,467	China
Blue mussel	200,070	Netherlands, France, Germany, Ireland, England
Mediterranean mussel	279,912	Spain, Italy
Other Mytilids (marine mussels)	569,166	China, Thailand, Philippines, New Zealand Korea, Malaysia, Chile
Brown seaweeds	1,914,972	China, Korea, Japan, North Korea, former USSR
Red seaweeds	862,630	Japan, Korea, Philippines, North Korea, Chile
Miscellaneous aquatic plants	209,753	Indonesia, China, Korea, North Korea, Japan

Over 1.5 million metric tons of fish and shellfish were harvested in Europe (including the former U.S.S.R.) in 1989. Six species–common carp, Mediterranean mussels, blue mussels, rainbow trout, Atlantic salmon, and Japanese oysters–accounted for about 86% of this total. The production of finfishes amounted to 886,750 metric tons, and 638,053 metric tons of mollusks. Table 9.10 summarizes European production.

Common carp and rainbow trout production in Europe will probably be stabilized at the current levels for the next few years. Atlantic salmon production increased rapidly during the late 1980s to about 200,000 metric tons in Europe by 1991, about five years earlier than was predicted. However, this rapid growth led to marketing difficulties, excess inventories, trade restrictions, and eventually to the $300 million bankruptcy of the government-legislated Norwegian Fish Farmer's Sales Organization in late 1991. In

TABLE 9.9. Aquaculture production of important marine species in Japan in 1989. (Source: FAO. 1991. Fisheries Circular 815 rev. 3).

Common Name	Scientific Name	Metric Tons
Yellowtail	*Seriola quinqueradiata*	153,164
Red seabream	*Pagrus major*	46,000
Black seabream	*Acanthopagrus schlegeli*	120
Coho salmon	*Oncorhynchus kisutch*	20,000
Japanese scad	*Trachurus japonicus*	6,000
Olive flounder	*Paralichthys olivaceus*	3,000
Puffers	Tetradontidae	1,150
Kuruma prawn	*Penaeus japonicus*	3,000
Japanese scallop	*Pecten yessoensiss*	180,255
Japanese oyster	*Crassostrea gigas*	256,313
Sea squirts	Ascidiacea	10,000
Japanese kelp	*Laminaria japonica*	65,000
Wakame	*Undaria pinnatifida*	108,451
Nori	*Porphyra tenera*	403,290

the beginning of the 1990s, the increase in European salmon production continued, as well as the rise in production in South America, especially Chile.[5]

The mussel crop (including both Mediterranean and blue mussels) is the most important cultured product by weight in Europe, constituting some 31% of all European aquaculture production. In close second place to mussels is the production of carps, especially in Eastern Europe. Carps constitute 30% of European aquaculture, with the majority (78%) being common carp. In contrast, oyster production accounts for only about 9.4% of the total.

In Europe, many finfishes are cultivated for fisheries enhancement and ranching to support both commercial and recreational

5. Source: "1992 Buyer's guide." *Seafood Leader* 12(2): 166.

TABLE 9.10. Aquaculture production of important species in Europe (including the former USSR) in 1989. (Source: FAO. 1991. Fisheries Circular 815 rev. 3).

Common Name	Scientific Name	Metric Tons
Common carp	Cyprinus carpio	360,735
Various other carps	Cyprinidae	98,888
Rainbow trout	Oncorhynchus mykiss	188,685
Various other trouts and chars	Salmo and Salvelinus	12,284
Atlantic salmon	Salmo salar	159,336
All other finfish		66,832
Total finfish		886,750
European flat oyster	Ostrea edulis	11,815
Pacific cupped oyster	Crassostrea gigas	131,696
Mediterranean mussel	Mytilus galloprovincialis	277,649
Blue mussel	Mytilus edulis	197,118
Various other mollusks		19,775
Total mollusks		638,053

fisheries. The stocking of Atlantic salmon smolts in rivers flowing into the Baltic Sea and into other sea areas in the eastern Atlantic Ocean is important, as is the stocking of coregonid species in the lakes of Eastern Europe and Finland. In addition to this, several salmonids including brown trout, char, and grayling are cultivated to support recreational fisheries.

Aquaculture production in Africa remains small, with a total yield of less than 95,000 metric tons. Nearly 97% of this is finfishes, and more than half the total is reported from a single country–Egypt. In Sub-Saharan Africa, aquaculture production is only about 36,000 metric tons and most (71%) is from Nigeria. Pond culture of tilapias and carps has been the primary type of culture. Small experimental and pilot projects for cultivation of African catfishes, freshwater prawns, and oysters are underway.

In the Near East there is aquacultural production of carps, tilapias, mullets, and other species. The leading country, Israel, produces nearly half (47%) of its own total fishery yields by aquaculture, primarily by using intensive polyculture. Total production in the Near East is about 38,000 metric tons annually.

Aquaculture production is growing rapidly in the U.S.A. Production in 1989 was 443,359 metric tons, or 3.16% of the world's total. Dramatic increases have come from catfish and freshwater crayfish, which amounted to almost 183,000 and 30,000 metric tons, respectively. Rainbow trout (26,000 metric tons), oysters (American and Pacific, 110,000 metric tons), bait minnows (8,500 metric tons), and ornamental fishes ($33 million) have been relatively stable but important aquacultural enterprises. Pacific salmon culture has been increasing significantly, climbing from 34,000 metric tons in 1986 to 73,000 in 1989. In addition, untold millions of Pacific and Atlantic salmon smolts have been released for fisheries enhancement projects. Over 844 million fingerling fishes, including salmons and various warmwater and coldwater sport fishes, were cultured and released into U.S. waters in one year (1980) by federal and state fisheries agencies.

Aquaculture production in Latin America is on the order of 242,000 metric tons per year. In addition, some 170,000 metric tons of finfish are derived from culture-supported fisheries in Brazilian, Cuban, and Mexican reservoirs. The herbivorous species "cachama" and "pacu" (*Colossoma* spp.) in Venezuela, Colombia, and Brazil are considered to have a great potential in Latin America and elsewhere. There is considerable interest and knowledge of shrimp culture in the region. Ecuador is one of the leading countries in the world in production of cultured penaeid shrimps, and production grew from 31,000 metric tons in 1986 to 72,000 in 1989.

Chapter 10

A Comparison of Aquaculture and Traditional Agriculture

INTRODUCTION

There are four principal methods of food production. Two are land-based: hunting and gathering, and agriculture; and two are aquatic-based: fishing and aquaculture. All four methods use basic ecological principles involving the flow of energy and matter. However, agriculture and aquaculture systems are human-made, artificial ecosystems that are not always based on the normal operation of the three fundamental ecological processes of production, consumption, and decomposition. Human cultivation of plants and animals for food tends to distort natural ecosystem dynamics. This is also true of hunting/gathering and fishing. Nevertheless, all food is produced in and by ecological systems that must respond to the basic laws of thermodynamics. In order to augment natural harvests, we must interfere with and change natural flows of energy and matter. Comparisons between production on land and in water and between yields in the fisheries and aquaculture are discussed in this section.

FOOD PRODUCTION ON LAND AND IN WATER

Facts and Figures

Data on international food production are available from the Food and Agriculture Organization (FAO). Production on land consists of plants (cereals, root crops, pulses, fruits, vegetables, beans, seeds, and sugar), and animals, including such products as milk and eggs. Aquatic production consists of shellfish, fish, marine mammals, invertebrates, and macroalgae. The figures presented in Table 10.1 show gross production. However, they do not indicate the real quantity of food available for direct human use. From a nutritional standpoint, it is important to calculate:

1. the edible part of food after processing;
2. the amount of available protein, fat, carbohydrate, vitamins, etc., in each food product; and
3. the nutritional characteristics of the particular food, for example, the composition of amino acids in proteins and fatty acids in lipids.

In addition, one should consider the amounts of cereals and fish that are utilized as animal feeds. A total of 3.65 billion metric tons of food items was produced in 1983. Of this gross production, 79.5% was of plant origin and 18.4% was of animal origin from land-based agriculture. This means that only 2% of the food by weight came from freshwater and marine areas.

Much of this food is not used for human nutrition, but is given to animals that produce meat, eggs, and milk. In the industrial world, about 60% of the cereals produced are fed to cattle, pigs, and chickens. For the world as a whole the figure is about 40%. About 25% of the total fish and shellfish harvests are transformed into

TABLE 10.1. Summary of the gross production of plant and animal products from land and water in 1983.

Total Annual Production	1000 metric tons	%
Terrestrial plants	2,900,000	79.5
Terrestrial animals	670,000	18.4
Aquatic animals	77,000	2.1
Aquatic plants	3,000	0.08
Grand total	3,650,000	

animal feed, mainly fish meals. Much of this is fed to pigs and chickens, but a significant quantity is incorporated into pelleted feeds for aquatic species.

We can adjust the figures to take account of total food production diverted to animal (mainly terrestrial animal) feeds. If we reduce the total amount of cereals produced by 40% and the amount of fish and shellfish by 25%, we obtain a true figure for food production available for human consumption (Table 10.2). For the sake of simplicity, only production of cereals and fish are adjusted and not that of terrestrial animals or aquatic plants, which are used very little in animal feeds. The total quantity of food available directly to humans, then, is close to 3 billion metric tons, of which 75% is derived from plants and 25% from animals. With a world population of about five billion people, the total amount of edible food products available for each person is about 600 kg per year.

The edible part of different food items varies considerably. Using tables published by the Swedish National Food Administration, it is possible to calculate the edible part of each food item by subtracting the amount of wastes from the gross quantity and thus obtaining a net quantity. The figures presented in Table 10.3 show that the net quantity of food available for human consumption after processing is about 1.8 billion metric tons or about 366 kg per capita. From this net quantity of food, it is then possible to calculate the quantity of protein, fat, and carbohydrate available. Protein is the least avail-

TABLE 10.2. Amount of food available for human consumption when 40% of the cereal production and 25% of the fish production is diverted for animal feed.

Adjusted Annual Production	1000 metric tons	%
Terrestrial plants	2,243,000	75.4
Terrestrial animals	670,000	22.5
Aquatic animals	58,000	2.0
Aquatic plants	3,000	0.001
Grand total	2,974,000	

able and therefore the most valuable component in human food. International availability is apparent in data presented in Table 10.4. The total amount of protein available for human consumption in 1983 was 156 million metric tons or about 33 kg per capita. Unfortunately, much food is wasted as a consequence of improper storage. In addition, in some parts of the world, many people eat much more than the minimum daily requirement. Finally, and even more importantly, is the fact that the human population in 1993 is no longer 4.7 billion but over 5.5 billion people.

Using a different approach, the FAO calculated the world annual per capita protein supply that reaches consumers to be only 27 kg in 1989. The sources of protein in the FAO determination were 62% plants, 33.5% terrestrial animals, and 4.5% fish and shellfish. Com-

TABLE 10.3. World production of food from land and aquatic areas available for human consumption after subtraction of production diverted to animal feeds after processing.

Net Annual Production	1000 metric tons	%
Plants	1,186,000	64.7
Terrestrial animals	619,000	33.8
Aquatic animals	27,000	1.5
Grand total	1,832,000	100

TABLE 10.4. World production of protein from land and aquatic areas. Net quantity after processing and subtraction of the quantities used as animal feed.

Net Annual Production	1000 metric tons protein	% of total protein
Plants	92,086	59
Terrestrial animals	59,000	38
Aquatic animals	4,900	3
Grand total	155,986	100

bining the two calculations gives a world per capita protein supply of 30 ± 3 kg, that is composed of 60.5% plant material, 35.7% terrestrial animals, and 3.8% fish and shellfish.

While the total amount of protein is critical, it is also important to look at the source of protein. This is because humans can utilize animal protein much more efficiently than plant protein, and some of the essential amino acids are in short supply in plant protein. Therefore, one cannot directly compare the amount of proteins available from plants and animals. In any case, fortunately, a mixture of both can satisfy all of a person's amino acid needs.

According to FAO, a 70 kg (154 lb) male adult requires the equivalent of 41 g of a reference protein (egg albumin) per day. According to Table 10.5, 63,900,000 metric tons of animal protein were produced in 1983. This will, theoretically, provide 4.3 billion people with their minimum annual protein requirements.

Table 10.3 showed the quantity of food produced after the amounts fed to animals are subtracted. The figures showed that only 1% of human food comes from aquatic animals and about 25% from animal production on land. That comparison alone, however, is misleading. We may obtain a more complete picture by way of three additional comparisons. Table 10.4 shows the total production of protein available for human consumption. Table 10.5 shows the animal protein production including milk and eggs. And finally, Table 10.6 shows the production of terrestrial meat protein alone compared with fish and shellfish protein production.

TABLE 10.5. World production figures for animal protein from land and aquatic areas—net quantity after processing and subtraction of quantities used for animal feed.

Net Annual Production	1000 metric tons	%
Meat, milk, eggs	59,000	92.3
Fish and shellfish	4,900	7.7
Grand total	63,900	100

It is therefore, important to make distinctions among gross food production, net food production, and finally, dairy, meat, and fish protein production. Only 1% of human food requirements are produced in aquatic areas when the net quantity of food available from all sources is considered. However, considering protein alone, some 2 to 3% is derived from aquatic animals; of animal protein, 8% is derived from fish and shellfish species; and if one simply compares meat and fish products, about 20% of human animal protein needs are supplied by fish and shellfish.

Of course, because aquaculture production accounts for only 15% (1990) of the total aquatic yield (excluding seaweeds), it contributes a proportionally smaller share of the world's protein than indicated by the figures above. Within a given region, however, the share of cultured aquatic protein may be much higher. In Asia, the production of aquatic animals through aquaculture in 1983 was about 1.6 kg per capita, compared to 5.5 kg per capita from harvest of wild stock; in other words, close to 30% of the total aquatic yield. The Philippines was the highest with 5.5 kg of fish and shellfish per capita, and Japan second at 2.8 kg.

The FAO is currently predicting an average annual increase of 5.5% for aquaculture and only 0.3% for natural fisheries from now until the year 2000. This means that by the end of this century, aquacultural production will account for 20 to 25% of the total aquatic yield–a significant increase in its importance for human nutrition.

TABLE 10.6. The world production of animal protein excluding milk and eggs–net quantity after processing and subtraction of quantities used for animal feed.

Net Annual Production	1000 metric tons	%
Meat	20,600	80.8
Fish and shellfish	4,900	19.2
Grand total	25,500	100

Reasons for Discrepancies
Between Land- and Aquatic-Based Production

Water is a formidable environment to terrestrial organisms such as ourselves; therefore, the work of cultivation is much more difficult there than on land, and great expenditures of effort are required. Construction of even simple containment structures outside natural water bodies is expensive and consumes much energy. Furthermore, cereal and other starch crops that provide the bulk of the calories in the human diet simply cannot be cultivated in aquatic environments and no equivalent aquatic crops (except, perhaps, for semiaquatic rice) have yet been cultivated on a mass, energy-efficient basis. It is possible to mass-produce unicellular algae and some nutritious vascular aquatic plants like the duckweeds, but the costs of collecting and dehydrating (water amounts of 96 to 99% of the biomass) still makes such mass production impractical.

Currently, therefore, most of the cultivated aquatic species are relatively high-value, which justifies the costs of their production. In other words, they generally constitute luxury rather than subsistence foods. This is true even in the People's Republic of China where tremendous quantities of carps are cultivated in very efficient systems integrated with conventional, terrestrial agriculture.

Chapter 11

Factors Promoting and Constraining Aquaculture

INTRODUCTION

Aquaculture has been looked upon as an excellent source of animal protein, a potential source of employment, a means to diminish trade deficits, and an instrument to utilize land and water resources more effectively. The prospects for aquaculture vary in different parts of the world depending on physical environments, markets, political attitudes, and socio-economic conditions. Factors that encourage aquacultural development are presented in Figure 11.1.

INCENTIVES FOR AQUACULTURAL DEVELOPMENT

Incentives for aquacultural development include a good market for the products, skillful entrepreneurs whose aim is to secure profits, and a suitable physical environment. A chain is no stronger than its weakest link. If any one of these factors is missing, there is no good reason to initiate an enterprise. The most important single factor when establishing a new company, regardless of its product, is a free market with a minimum of regulations and without constraints on trade.

ESTABLISHING AN AQUACULTURAL ENDEAVOR

The development of an aquacultural endeavor may be obstructed or prevented if several additional prerequisites are not present. First, legislation must be favorable or readily modifiable. Here, both environmental and general agricultural and fisheries legislation must be considered. In fact, if the political will to create an aquacultural endeavor is missing and/or other competing governmental agencies are very strong in opposing aquaculture, whether on the national or any other level, little significant aquacultural development can be expected.

Another factor of great importance is institutional support. A new aquacultural industry is dependent upon support from public institutions, including universities. Governmental hatcheries serve as repositories for important gene pools and in some cases, provide aquacultural enterprises with initial seed. These hatcheries are also sources of technology, including methods already developed for local species and those later refined in response to industry needs. Even more important, at certain times, are institutions that provide disease diagnosis and treatment recommendations. Health care and quarantine programs are essential. Most individual aquacultural endeavors simply do not have the resources or personnel to diagnose diseases and determine appropriate treatments.

The environmental impact of aquaculture must be considered in order to prevent adverse impacts on waters receiving their effluents. The nature of effluents, the means to reduce pollutants, and the carrying capacity of receiving waters must be assessed. Aquaculture industries, again, generally do not have the resources necessary to do this, and require governmental assistance. Decisions on issuing licenses must be based on fair-minded scientific assessments of the relationships between aquaculture and the environment.

Technology is the basis for the development of any new industry, and new technology must be developed as conditions change if any industry is to remain competitive. Aquacultural equipment is continuously becoming more sophisticated. Demands for better cages, nets, anti-foulants, feeds, feeders, therapeutic and prophylactic drugs, aeration devices, and other equipment are driving forces that determine continued success or ultimate failure of an aquacultural endeavor. Fortunately, as such an industry matures, much of the need for governmental research and development is transferred to the private support industry that invariably develops around the primary industry.

FIGURE 11.1. Factors that encourage aquaculture development. Source: Ackefors 1986.

A. NATURAL PREREQUISITES

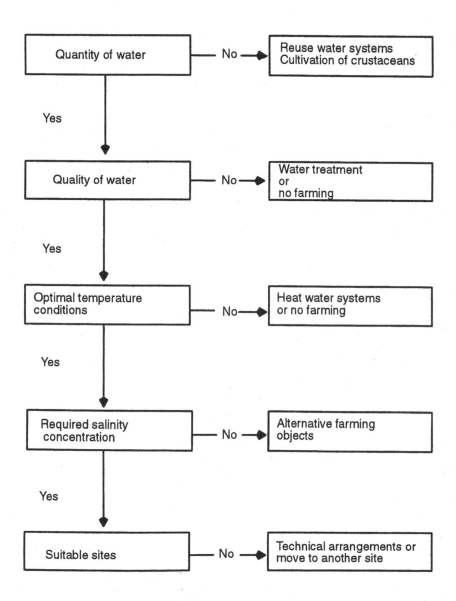

FIGURE 11.1 (continued)

B. BIOLOGICAL AND TECHNICAL OPTIMIZATION

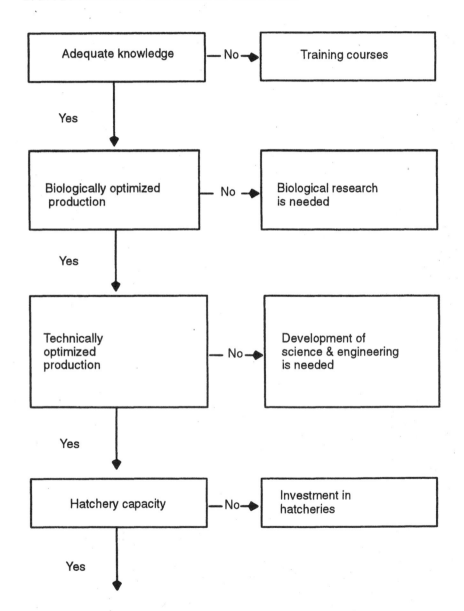

C. MARKETING

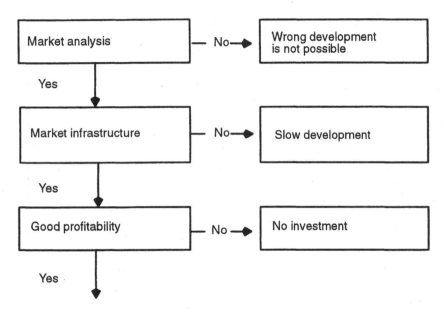

D. LEGAL REGULATION AND ENVIRONMENTAL IMPACT BY AQUACUL-TURE

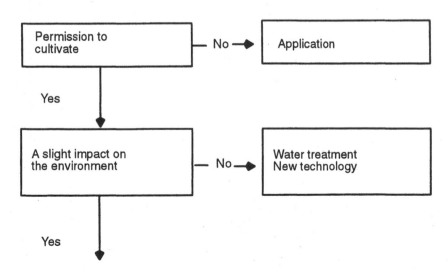

THE OPERATION OF AN AQUACULTURAL ENTERPRISE

The operation of an aquacultural enterprise is dependent basically on capital, seed and feed, and competent management. The operating costs are usually a serious problem during the start-up phase of a newly established company. The firm or individual needs a loan to cover not only construction costs but also operating costs for the first two to three years until the first or second harvest generates a positive cash flow. This is an obstacle for all new aquacultural enterprises around the world, from small cottage industry farms to huge vertically integrated firms.

High quality seed is a crucial factor. A governmental program for breeding and providing such seed may help, but most aquacultural enterprises are forced to purchase whatever seed is commercially available. Seed source and quality must be considered carefully before any capital is committed to a new venture. Once acceptable seed is secured, it must be provided with nutritional, reasonably priced feed(s). This is crucial because feed costs account for 40 to 60% of operating costs in most finfish and crustacean aquacultural enterprises.

Competent management is critical to the operation of an aquacultural enterprise, but its role is often under-emphasized or ignored. The operation of a fish or shrimp farm is time-intensive, and usually requires attention around the clock. The successful farm manager must be an energetic jack-of-all-trades: able to organize and motivate farm workers, recognize and correct biological and water chemistry problems, and perhaps most important, be able to make effective emergency field repairs to complex equipment as a matter of routine. In addition, the manager must, at times, use skills usually associated with a purchasing agent, diesel mechanic, heavy equipment operator, and hydrological engineer. The fact that this mix of talents is required is common knowledge to farmers and agriculturists, who operate this way all the time, but it is not obvious to the biologists and entrepreneurs who are often attracted to aquaculture.

MARKETING OF AQUACULTURAL PRODUCTS

In a growing industry such as aquaculture, marketing one's products through commercial channels is a must. At first, an individual aquaculturist may be able to sell the product directly to consumers or local outlets, but will soon be trapped in a precarious position as competition develops. At that point, a broker or cooperative of producers becomes necessary to handle sales, processing, and distribution.

In the competitive world of commercial seafood sales, successful dealers generally offer a variety of products, including value-added ones that may have been smoked, filleted, frozen, preportioned, preseasoned, etc. In addition, these products are strongly promoted through advertisements to both the general public and to the restaurant, seafood, and grocery industries in their own specialized publications. Most individual aquaculturists would prefer to concentrate on production, and in any case, individual producers are usually unable to undertake the comprehensive processing and promoting activities required for success. For them, a cooperative enterprise that can employ seafood marketing professionals may be the solution.

Because aquacultural endeavors are often started in areas of low population density, they may soon become dependent both on more distant domestic and on foreign export markets. Without a strong sales organization, it is difficult to distribute products far from the point of production. Furthermore, there are critical considerations such as transport/import regulations and quality control standards which must be met. These are especially stringent for products destined for human consumption (which includes most aquacultural products). Again, these activities are performed more efficiently by marketing the product through either a broker or a cooperative.

FACTORS CONSTRAINING AQUACULTURAL DEVELOPMENT

The factors required for promoting aquacultural development are also constraints. Those who develop aquacultural ventures will certainly fail if they are overly optimistic or overlook these requirements.

Aquaculture is a young business in most of the industrialized world. True, it can trace its roots back about 4,000 years; nevertheless, in sharp contrast with the 10,000-year history of traditional agriculture, much of the aquaculture industry has developed only during the past three decades. Thus, there remains a serious lack of knowledge and experience when it comes to the culture of many aquatic species. Entrepreneurs tend to minimize biological, technical, and economical problems in scaling up projects from experimental production units to commercial units. Biologists and fisheries personnel may minimize the business, managerial, financial, and marketing problems of a commercial-sized venture. There is generally a lack of training at all levels from the farmer to the scientist. Agricultural extension services in many places are unable to assist developing aquacultural ventures due to a lack of extension personnel with specialized training in aquatic biology and aquaculture. Finally, where adequate technical resources are available to aquaculturists, they may still be constrained by competition for land and water.

Chapter 12

Conclusions and Prospects for Future Development

The prospects for aquaculture production are very good but there are some constraints that must be addressed. Aquaculture remains a relatively young form of business in commercial terms. A research base is only now emerging to assist the industry. As in many developing endeavors, there is a serious lack of entrepreneurial knowledge and experience in aquaculture. Entrepreneurs have often minimized very real biological, technical, and economical problems and this has led to the failure of well-funded and publicized projects. The lack of training at all levels from farmers to extension advisors to scientists is obvious. In some locations there is legal, political, and/or environmental opposition to aquaculture development. Even in areas where sufficient knowledge and support services are available to permit development of profitable aquacultural endeavors, competition for land and water may become a major obstacle to limit further development of aquaculture.

But despite these gloomy reflections, aquaculturists will likely overcome these difficulties and the industry will develop to the point where it supplies the majority of the world's fish and shellfish. The most significant driving force for this development will be the expanding world population that will lead to increasing demand for aquatic products coupled with the simultaneous decreasing ability of traditional fisheries to meet the demand. In addition, aquaculture has advantages over other traditional forms of terrestrial animal husbandry in being more efficient in terms of land, feed, and energy.

The basic elements in aquaculture production are land, water, seed, and feed. In addition, energy must be added directly and/or indirectly to the production process. Depending on the methods and technology involved, energy costs may vary considerably. All these factors, then, combine to determine product quality and its final value. Extensive aquaculture systems often use natural waters, seed, and feed, so their costs may be negligible. Such aquacultural methods usually require large areas and low stocking densities. Although intensive aquaculture techniques will probably become the dominant source of food species, extensive methods, often integrated with traditional agriculture, will play an important role, especially in developing countries. In addition, the cultivation of marine molluscan and algal species will remain largely extensive in nature, although there should be some movement toward use of semi-intensive methods.

Fisheries enhancement, the release of juvenile fishes and shellfishes in open water, is a major goal of many nations. Enhancement can support either commercial or recreational fisheries. Ocean ranching, in which the released young of migratory species eventually return to the culturist after a period of growth in open water, is a type of aquaculture that is likely to assume greater importance in the future as the demands on natural fisheries increase.

The amount of water needed may vary considerably depending on the system. The opposite extreme is intensive aquaculture, which requires relatively small cultural units, high stocking densities, usually with significant inputs of expensive feeds, and significant amounts of water per unit of production. Water quality is also of immediate concern to assure that the species will survive in the cultural units. Temperature, oxygen, nitrogen compounds, carbon dioxide, pH, and alkalinity/hardness are crucial water quality parameters.

Less energy is used for producing fish than for most terrestrial livestock, especially swine and beef. Fish do not have to maintain a constant body temperature and the energy required to maintain position and to move in the water is much less than that required by terrestrial mammals and birds. Fish excrete most of their nitrogenous wastes as ammonia rather than as urea or uric acid. Therefore, they expend less energy in protein catabolism and excretion of nitrogenous wastes. The expenditure of energy for producing chicken (one of the most efficient terrestrial animals) is generally comparable to that for finfish production. But the conversion rate of protein into flesh is higher in fish than in chicken. Some forms of aquaculture, such as the current culture practices for shrimp and prawns, require a higher energy input, comparable to that required for beef production. But these high energy methods are at least partly a result of high shrimp demands and prices. Even with shrimp culture, lower intensity, energy efficient methods are available.

Water and energy are limiting factors for production of all food and industrial goods. Water is also usually required for the elimination of wastes, which is a third limiting factor. There is a limit to the waste-carrying capacity of a water body, and the environmental impact of aquacultural endeavors on the world's waters will be decisive in determining the growth of aquaculture. The release of nutrients from aquacultural enterprises is a problem, given the current level of waste treatment technology. Even in the marine environment, there is a trend to move cage culture units away from the coastal zone to minimize the effects of nutrients on coastal waters. Large cage systems designed for open sea operation are now being put into use.

The treatment of water discharged from land-based aquaculture systems can be very expensive, and state-of-the-art methods often leave much to be desired. This is largely due to the fact that great amounts of water with a very low concentration of nutrients are being discharged. Quantity per unit volume is low but absolute quantity can be very high because of the water volumes involved. The most likely solution to this dilemma will be the increasing reuse of water on fish farms, with corresponding reduction in total discharge.

Aquaculture products will become more widely distributed. Although more than 500 aquatic species are now being cultivated, and many are of great local or regional importance, very few are of major importance on a global scale. Those few species that are international in distribution, either naturally or as a consequence of translocations, include tilapias, various Chinese carps, salmonids, oysters of the genus *Crassostrea*, mussels of the genus *Mytilus*, and a crayfish, *Procambarus clarkii*.

In addition, the rainbow trout and the common carp are cultivated on all continents except Antarctica. They are reared in more countries than any other species and are currently considered to be the only real domesticated aquatic species cultured for food, that is, the only species that have been cultured long enough to allow the development of genetic strains that are suited to farming and are distinct from wild strains. Domestic strains are usually more efficient and economic than wild ones on the farm and less efficient and viable in the wild. When nonnative species are introduced into a region, it is preferable to use domesticated strains that may not be competitive in the wild, and therefore less environmentally threatening should some escape.

The number of aquatic species cultivated and eventually domesticated will increase as new technology becomes available; but progress in this regard is dependent on research. For example, the rearing of marine fishes and crustaceans, which, in principle, is much more difficult than the rearing of freshwater and anadromous species, is most developed in Japan, where there is a national commitment to develop aquaculture. Over 3,000 Japanese scientists are engaged in aquaculture-related research. The first step in domestication is closing the life cycle, that is, being able to propagate the species in captivity and then raise the young to maturity and repeat the cycle. Once culture is closed and independent of wild stocks as a source of either young or mature adults, then genetic selection can proceed and a true domesticated strain can be produced.

An international shortage of high-quality fish, shellfish, and marine macroalgae has created demand for high-quality aquaculture products and has enabled profitable aquacultural endeavors to develop. This favorable market situation should continue to drive further aquacultural development, with industrialized countries emphasizing high-value products and developing countries producing both high-value products for export earnings and lower-value products for domestic consumption.

The aquaculture industry is growing rapidly in many countries—in some, as much as 10 to 40% annually. The projected annual increases will be a more modest 5% through the year 2000. Aquaculture production should approach 20 million metric tons by then. Some 80% of this will be produced in Asia with Europe accounting for 12%. Aquaculture will then generate 20 to 25% of all aquatic products. Aquaculture will continue to grow because there will continue to be an increased world-wide demand for high-quality, healthful seafood. Traditional fisheries, already near maximum production capacity (and some would say over maximum), will be increasingly constrained by competing uses such as recreational activities, conservation of endangered marine animals (e.g., turtles and porpoises), and industrialization with its resulting pollution and contamination. These constraints alone would cause seafood markets to become increasingly dependent on aquaculture. But, in addition, the world population currently at about 5.5 billion will likely double in just 40 years. Only aquaculture can supply the increased fishery needs of the next century.

Further Readings

Ackefors, H. and M. Enell, 1993. Pollution loads from landbased and water-based systems. Journal of Applied Fisheries (in press).

Ackefors, H. and M. Enell, 1990. Discharge of Nutrients from Swedish Fish Farming to Adjacent Sea Areas. AMBIO, Vol., 19(1): 28-35.

Ackefors, H., 1986. Prospects and limitations for Aquaculture in Scandinavia. Inst. Freshwater Research, Drottningholm. Rep. No. 63: 5-16.

Ackefors, H. and C.-G. Rosén, 1979. Farming Aquatic Animals. The Emergence of a World-Wide Industry with Profound Ecological Consequences. AMBIO, Vol. 8(4): 132-143.

ADCP. 1984. *Inland Aquaculture Engineering. Lectures presented at the ADCP Inter-regional Training Course in Inland Aquaculture Engineering.* ADCP/REP/84/21, UN/FAO. 591 pp.

Aquacultural Engineering, Amsterdam: Elsevier.

Aquaculture, Amsterdam: Elsevier.

Bardach, J. 1985. "The role of aquaculture in human nutrition." *GeoJournal* 10(3): 221-232.

Bardach, J. E., J. H. Ryther, and W. O. McLarney. 1972. *Aquaculture, The Farming and Husbandry of Freshwater and Marine Organisms.* New York: Wiley-Interscience. 868 pp. ISBN 0-471-04826-7.

Barnabé, G. 1990. *Aquaculture.* Vol. 1, 528 pp., Vol. 2, 584 pp. Chichester: Ellis Horwood Ltd., Transl. by L. M. Laird.

Barnabé, G. (ed.). 1986. *Aquaculture.* Vol. 1-2. Technique et Documentation Paris: Lavoisier. 1123 pp. ISBN 2-85206-302-6.

Bilio, M., H. Rosenthal, and C. J. Sinderman (eds.). 1986. *Realism in Aquaculture: Achievements, Constraints, Perspectives.* Bredene, Belgium: European Aquaculture Society. 585 pp. ISBN 90-71625-01-X.

Boyd, C. E. 1984 (3rd printing). *Water Quality in Warmwater Fish Ponds.* Auburn University Agricultural Experiment Station, Auburn, AL, U.S.A. 359 pp.

Csavas, I. 1992. "Recent developments and issues in aquaculture in Asia and the Pacific." Paper presented at the APO Seminar on Aquaculture, August 25 - September 4, 1992. Tokyo.

Dupree, H. K. and J. V. Huner (eds.). *Third Report to the Fish Farmers. The Status of Warmwater Fish Farming and Progress in Fish Farming Research.* U.S. Fish & Wildlife Service, Washington, DC, U.S.A. 270 pp. 1984.

EIFAC. 1986. *Flow-through and Recirculating Systems. Report of the Working Group on Terminology, Format and Units of Measurements.* EIFAC, Technical Paper 49. Rome. ISBN 92-5-102416-2.

European Aquaculture Society. 1976-1990. Quarterly Newsletters No. 1- , Bredene: Belgium.

European Aquaculture Society. 1990. Aquaculture Europe, Magazine of the European Aquaculture Society. Vol. 16(1), Bredene: Belgium.

FAO/UN. Fisheries Technical Papers, Rome.

FAO/UN. 1992. Fisheries Circular No. 815 revision 4. *Aquaculture Production, 1984-1990,* Rome.

FAO/UN. 1991. Fisheries Circular No. 815 revision 3. *Aquaculture Production, 1984-1989,* Rome.

FAO/UN. 1991. *Fishery Statistics: Catches and Landings 1989.* Vol 68. FAO Fisheries Series No. 36, Rome.

FAO/UN. 1988. *Topography for Freshwater Fish Culture: Topographical Tools.* FAO Training Series, No. 16/1, Rome.

FAO, 1988-89. A global overview of Aquaculture based on regional surveys:

ADCP/REP/88/30 A Regional Survey of the Aquaculture Sector in Eleven Middle East Countries

ADCP/REP88/31 A Regional Survey of the Aquaculture Sector in East Asia

ADCP/REP/88/32 A Regional Survey of the Aquaculture Sector in the Pacific

ADCP/REP/89/33 Planning for Aquaculture Development–Report of an Expert Consultation held in Policoro, Italy, 26 July–2 August 1988

ADCP/REP/89/34 A Regional Survey of the Aquaculture Sector in the Mediterranean Region

ADCP/REP/89/35 A Regional Survey of the Aquaculture Sector in West Asia

ADCP/REP/89/36 A Regional Survey of the Aquaculture Sector in Africa, South of the Sahara

ADCP/REP/89/37 A Regional Survey of the Aquaculture Sector in North America

ADCP/REP/38 A Regional Survey of the Aquaculture Sector in Eastern and Northeastern Europe

Grimaldi, E. and H. Rosenthal (eds.). 1988. *Efficiency in aquaculture production: Disease control.* Proceedings of the Third International Conference on Aquafarming "Acquacoltura 86" held in Verona, Italy, October 9-10, 1986. Milano, Edizione del Sole 24 Ore, 226 + 227 p.

Grimaldi, E. and H. Rosenthal (eds.). 1986. *Trends and Problems in Aquaculture Development.* Proc. 2nd Int. Conf. Aquafarming, Ente Autonoma Fiera di Verona, Verona: Italy.

Halver, J. E. (ed.) 1989. *Fish Nutrition.* New York: Academic Press.

Halver, J. E. and K. Tiews (eds.). 1979. "Finfish nutrition and fishfeed technology." *Schr. Bundesforschungsanst. Fisch., Hamb.,* (14/15) Vol.1: 593 p., Vol. 2: 622 p., Berlin.

Huet, M. 1972. *Textbook of Fish Culture.* Fishing News Books, Ltd., Surrey: England. 436 pp. (Translated by Henry Kahn from the Fourth French Edition of Trait de Pisciculture, published by Editions Ch. De Wyngaert, Brussels, 1970.)

Huisman, E. A. (ed.). 1986. Aquaculture research in the African Region. Proceedings of the African Seminar on Aquaculture organised by the International Foundation for Science (IFS), Stockholm, Sweden, held in Kisumu, Kenya, 7-11 October 1985. Wageningen, Netherlands, PUDOC, 274 p.

Huner, J. V., 1993 (ed.). *Freshwater Crayfish Aquaculture in North America, Europe, and Australia: Families Astacidae, Cambaridae, and Parastacidae.* The Haworth Press, Inc. (In print).

Journal of Aquaculture in the Tropics. Calcutta: Oxford and IBH Publishing Co. PV. T. Ltd.

Korringa, P. 1976. *Farming Marine Fishes and Shrimps.* Amsterdam: Elsevier, 208 pp. ISBN 0-444-41335-9.

Korringa, P. 1976. *Farming the Flat Oysters of the Genus* Ostrea. Amsterdam: Elsevier, 238 pp. ISBN 0-444-41334-0.

Korringa, P. 1973. *Farming Marine Organisms Low in the Food Chain.* Amsterdam: Elsevier, 264 pp. ISBN 0-444-41332-4.

Liao, I. 1988. "East meets West: An Eastern perspective of aquaculture." *Journal of the World Aquaculture Society* 19(2): 62-73.

Nash, C. E. 1988. "A global overview of aquaculture production." *Journal of the World Aquaculture Society* 19(2): 51-58.

New, M. B. 1991. "Turn of the millennium aquaculture." *World Aquaculture* 22(3): 28-49.

Pillay, T. V. R. and W. A. Bill (eds.). 1979. "Advances in Aquaculture." Papers presented at the FAO Technical Conference on Aquaculture, Kyoto, Japan, May 26 - June 2, 1976.

The Progressive Fish-Culturist. American Fisheries Society, 5410 Grosvenor Lane, Bethesda, MD. U.S.A. 20814.

Pullin, R. S. V. and R. H. Lowe-McConnel (eds.). 1982. *The Biology and Culture of Tilapias.* International Center for Living Aquatic Management, Manila: Philippines, 432 pp. ISBN 971-04-0004-5.

Sandifer, P. A. 1988. "Aquaculture in the West, a perspective." *Journal of the World Aquaculture Society* 19(2): 73-84.

Shang, Y. C. 1981. *Aquaculture Economics: Basic Concepts and Methods of Analysis.* Boulder: West View Press, 153 pp. ISBN 0-7099-2318-X.

Wheaton, F. W. 1977. *Aquacultural Engineering.* New York: Wiley Interscience, 708 pp. ISBN 0-89874-788-0.

World Aquaculture Society, Journal of, 1986-present, Baton Rouge.

World Mariculture Society, Journal of. 1979-1985, Baton Rouge.

World Mariculture Society, Proceedings of, 1970-1978, Baton Rouge.

SEMI-TECHNICAL AQUACULTURAL PUBLICATIONS

Aquaculture Magazine–P.O. Box 2329, Asheville, NC 28802, U.S.A.

Fish Farming International–Meed House, 21 John Street, London WC1N 2BP, Great Britain.

NAGA, The ICLARM Quarterly Publication, issued by ICLARM, Manila, the Philippines.

World Aquaculture Magazine, 1988-present, published by the World Aquaculture Society, 16 East Fraternity Lane, Baton Rouge, LA 70803, U.S.A.

Glossary of Aquaculture Terms

Algae. Unicellular or multicellular so-called cryptograms which live mainly in water. They include several larger groups such as red algae, brown algae, green algae, and diatoms. Larger macroscopic algae are usually called seaweeds. Small microscopic algae that drift in the water are called phytoplankton. Representatives of both groups are cultivated.

Alkalinity. The combined effect of mineral substances measured by the power of a solution to neutralize hydrogen ions; usually expressed as an equivalent of calcium carbonate.

Anadromous Fish. Fish that mature in the sea but migrate to freshwater to reproduce. Atlantic and Pacific salmon, etc.

Aquaculture. The cultivation or farming of aquatic organisms such as fish, shellfish (mollusks and crustaceans), and algae using extensive or intensive methods to increase production or yield to a level above that naturally available in the environment. Farming also implies individual or corporate ownership of the stock being cultivated

BOD. Biological Oxygen Demand–a term describing the amount of oxygen consumed by respiratory processes in an aquatic environment. It is customarily measured as BOD_{96}, the amount of oxygen consumed by a confined sample during a period of 96 hours at $20°C$ in darkness–measured in mg/l. It is a gauge of the amount of decomposable organic material present.

Buffer. A substance or substances in solution that resist or counteract changes in pH that would otherwise result from the addition of acid or alkali to the solution.

Cage Culture. See Net Pen Farming.

Carnivore. Literally a meat-devourer; a consumer of or a feeder on other animal species.

Catadromous. Fish that mature in freshwater and migrate to the sea to reproduce. European eel, mullets, etc.

COD. Chemical Oxygen Demand–a term describing the amount of oxygen required to oxidize all organic matter in an aquatic environment. It is measured by oxidizing organic matter present in a sample with a strong oxidant (potassium dichromate)–measured in mg/l.

Compensatory Stocking. The stocking of juvenile fishes to compensate for damage to natural populations by industrial activities such as construction of hydroelectric plants.

Crustaceans. Mainly aquatic invertebrate animals with jointed appendages and flexible exoskeletons such as crayfish, lobsters, crabs, shrimps, etc.

Detritivore. Organism that consumes microbially enriched organic detritus.

Detritus. Decomposing organic matter.

Ecology. The study of the interactions of all living organisms and their environment, including both biotic and abiotic features.

Ethology. The study of animal behavior.

Extensive Aquaculture. That in which energy and supplemental inputs from man are minimal. Most food derives from natural food chains. Examples include low-density cultivation of fishes and shrimps in the tropics and high-density culture of bivalve mollusks in temperate climates.

Feed. Organic material supplied to and consumed directly by heterotrophic organisms. The various types of feed are dry feed, stabilized, semimoist feed, and wet feed. Dry feeds have low moisture levels and are usually provided in pelleted form. Semimoist feeds have intermediate moisture levels and contain stabilizing chemicals to prevent spoilage in storage. Wet feeds are ground fish and fish offal often combined with vitamin and mineral premises and binding agents. Storage life is short.

Feed Coefficient. The dry weight of feed required per unit live (wet) weight gain (feed/gain). It is a special form of feed conversion ratio.

Feed Conversion Ratio. Amount of feed supplied divided by weight gain. The same ratio might yield a different figure depending on whether wet, dry, or some combination of wet and dry weight units are used.

Feed Efficiency Ratio. The inverse of the feed conversion ratio (gain/feed).

Fingerling. Any juvenile fish from advanced fry to the age of one year from the date of hatching. The term is usually applied to fishes under 100g in weight, but this is not rigid and varies according to species.

Fisheries. Areas or the arrangements for the capture of fish and shellfish from natural waters using various techniques. Commercial fisheries are "for profit" endeavors. Sport fisheries are "for recreation" endeavors.

Fisheries Enhancement. The stocking of young fish/shellfish to increase the existing population densities.

Fisheries Management. Conservation of fish and shellfish populations in natural environments, usually to insure optimum sustainable yields. This is accomplished by adopting certain measures, such as the stocking of fish, the introduction of prey organisms, biotope maintenance (fertilization, liming, preparing spawning grounds, etc.) and the regulation of the catch.

Fish Farming. A common term for aquaculture.

Fry. Fish and shellfish larvae, the youngest juvenile stages. **Sac fry** are young fish from hatching to the time the yolk sac is absorbed. **Swim-up fry** have just absorbed the yolk sac, become buoyant and are ready to consume food.

Hardness. Theoretically, the concentration of all the metallic cations, except those of the alkali metals, present in a water; in general, for all practical purposes, hardness is a measure of the concentration of calcium and magnesium ions in a water. Frequently expressed as mg/l calcium carbonate equivalents.

Hatchery. A facility where fish and shellfish are spawned and juveniles ("seed") are produced.

Herbivore. Consumer or feeder on plant species.

Integrated Aquaculture. Farming aquatic and terrestrial organisms in one unit. Wastes and/or products from terrestrial agricultural production are used to cultivate fish and/or shellfish.

Intensive Aquaculture. That in which most food and external energy is supplied by man. The organisms cultured are concentrated in a small area. Cultivation is carried out in ponds, troughs, etc., on land or in net pens, cages, etc., in lakes, rivers, or coastal areas.

Macroalgae. Algae that are visible to the naked eye, often attached to the bottom or a substrate. Nonvascular seaweeds such as kelp, nori, dulse, etc.

Mariculture. Aquaculture in the marine environment.

Microalgae. Microscopic algae, usually unicellular, free-floating (phytoplankton), or attached to submerged substrates.

Mollusks. Diverse group of invertebrate animals usually with a hard shell composed of calcium carbonate except in cephalopods (squids and octopuses). Includes gastropods (snails, conchs, and abalones) and bivalves (oysters, scallops, mussels, clams, etc.).

Monoculture. The cultivation of one species only.

Net Pen Farming. The farming of fishes such as rainbow trout and Atlantic salmon in floating enclosures (net pens or cages) in lakes or coastal areas. The fishes are highly concentrated in the enclosures and must be fed a nutritionally complete diet. Wastes are removed and dissolved oxygen is supplied by the flow of water through the walls of the cage.

Nutrients. Compounds that stimulate plant growth in the aquatic environment. Phosphorus and nitrogen compounds are macronutrients.

Omnivore. A species that consumes plants and animals and, possibly, decomposers such as fungi and bacteria and detritus.

Person Equivalent (PE). An expression of the quantity of domestic sewerage produced daily by a normal human. The average PE values are 12 g nitrogen/day, 2.5 g phosphorus/day and 75 g oxygen/day (as BOD).

pH. A measure (the negative logarithm) of the hydrogen ion concentration in soil and water. A pH less than 7.0 is acidic; a pH of 7.0 is neutral; and a pH over 7.0 is alkaline, or basic.

Polyculture. The cultivation of several species together.

Population. A group of individuals belonging to the same species, living in a delineated location, and being actually or potentially related. It is possible to have more than one population of the same species of fish in the same lake. For example, if one group reproduced upstream and one group reproduced downstream, there would be two genetically distinct populations.

Predator. An animal that kills and eats other animals.

Processing. Conversion of fish, shellfish, and algae into various products by filleting, smoking, drying, salting, etc.

Protein Efficiency. The live weight gain per unit weight of protein consumed. Also called the protein efficiency ratio (PER).

Put-and-Take Fishing. The stocking of fish of a size suitable for fishing in recreational waters.

Raceway. Usually a long narrow culture pond (often made of concrete) with average dimensions of $25.0 \times 3.0 \times 1.0$ m, and with the water inlet and outlet at opposite ends, thus permitting fish to be grown in streams of controlled velocity.

Ranching. The stocking of juvenile migratory fish, such as salmonids, for ongrowing in the sea (ocean ranching) or in lakes (lake ranching). The fish are eventually caught when they return to the area where they were released.

Recirculating System. An intensive culture facility in which the water is recycled.

Reproduction. The production of offspring.

Seaweed. Certain species of macroalgae including red, brown, and green algae.

Secchi Disk. Black and white disk, 20 cm in diameter, which is lowered into a body of water to measure its transparency. Clear waters are usually nonproductive, while excessively turbid waters usually have too much SS and BOD. The ideal secchi disk transparency in fish ponds is 45 cm.

Seed. Eggs and juveniles of species (cultured or caught) stocked in aquaculture facilities or natural water bodies.

Settling. The transition made by certain mollusk larvae from a free-swimming stage to a stationary, settled stage.

Shellfish. Crustaceans and mollusks.

Smolt. Juvenile salmons and trouts that leave their freshwater nursery areas and migrate to the sea to mature.

Spat. Recently settled, juvenile oysters. Sometimes applied to other sessile bivalve mollusks.

SS. Suspended solids. The amount of particulate material in a water sample. It can be used to estimate Biochemical Oxygen Demand (BOD), because most of the BOD is usually a result of material found in the suspended solids.

Supplemental Stocking. The stocking of juvenile fish and shellfish to increase the population densities in natural waters.

Index

Page numbers followed by t indicate tables; page numbers in italics indicate figures or illustrations.

T - #0599 - 071024 - C0 - 210/140/9 - PB - 9780367401979 - Gloss Lamination